PETROCHEMICAL TECHNOLOGY ASSESSMENT

PETROCHEMICAL TECHNOLOGY ASSESSMENT

DALE F. RUDD
University of Wisconsin-Madison

SAEED FATHI-AFSHAR
ETH
Zurich, Switzerland

ANDRES A. TREVIÑO
ASPA S.A.
Monterrey, Mexico

MARK A. STADTHERR
University of Illinois-Urbana

A WILEY-INTERSCIENCE PUBLICATION
JOHN WILEY & SONS
New York • Chichester • Brisbane • Toronto

CHEMISTRY

6391- 895X

ISBN 0-471-08912-5

Printed in the United States of America

10 9 8 7 6 5 4 3 2 1

Foreword

The modern petrochemical industry evolved rapidly from its initial major stimulus during World War II. Dale Rudd and his associates in a series of studies showed that the industry has been characterized by the continual supersedure of existing processes and products by new and more efficient technologies. These replacement technologies also can have indirect impact on the supply and demand of feedstocks and by-products.

An excellent example of this phenomenon is the case of acetic acid manufacture. The early commercial procedure using ethylene has now been replaced by a more efficient process using methanol and synthesis gas. Thus not only has the feedstock been completely changed but the decreased demand for ethylene to meet acetic acid requirements has led to decreased production of co-products which are produced during the manufacture of ethylene.

In a technically advanced society, a new process must be evaluated not only from its competitive economics but also in regard to its potential impact in the complex interrelationship between industry and the marketplace. Professor Rudd and his associates have made an important contribution in recognizing that it is possible to express this relationship in the form of a linear model, which can be useful as a powerful planning tool for assessing the impact of chemical industry trends. However, in the development of the model, the data base and the techniques for estimating economics and relative impact of fuel, utility, and feedstocks require more extensive background and experience than is customarily available in a university environment.

In early discussions on this matter, we in Exxon Chemical Company recognized that this was an excellent opportunity to utilize information available within our organization which was not of a proprietary nature to help improve the utility of the model. It was an area in which university-industry cooperation would be constructive.

The Exxon Chemical ¬Company was pleased therefore to invite Professor Rudd as a consultant and two of his graduate students, A. Treviño and S. Fathi-Afshar, to spend time in residence in our research laboratories obtaining background for making process economic studies and to collect specific process data for use in strengthening the data base for the program. Exxon was also able to suggest several variations in the model which were useful in this regard.

Exxon is pleased to have been able to assist in these studies, which have led to the preparation of this publication and to making available a tool that can constructively further development of the petrochemical industry during what we perceive to be a period of accelerating change.

D. S. Maisel
Senior Planning Advisor
Exxon Chemical Company
Florham Park, N.J.
January 28, 1981

Preface

In 1972 a joint research project was initiated among government, industrial, and university organizations to analyze the performance of the chemical industry. The fundamental questions considered include:

1 What are the ultimate performance characteristics of the industry in terms of feedstock, energy, and investment capital requirements?

2 Is it possible to anticipate the effects on the industry of changing product lines, raw materials sources, process technology, and government legislation?

3 What are the characteristics of new processes and products that lead to their eventual adoption by the industry?

4 How is the pricing structure of the industry influenced by the forces of technology and the economy?

This work was performed at the University of Wisconsin, the Chemical Technology Department of Exxon Chemical Company, and Shanahan Valley Associates. A special acknowledgment is given to Dr. J. F. Mathis, Mr. W. W. Yuan, Dr. D. S. Maisel, and Mr. S. D. Goldman of Exxon Chemical Company.

The major contributors to this project include The University of Wisconsin-Madison, The National Science Foundation, Exxon Chemical Company, D. C. Slichter Professorship, Shanahan Valley Associates, The American Chemical Society, Midwest Universities Consortium for International Activities, Alfa Industrias (Monterrey, Mexico), Consejo Nacional de Ciencia y Technologia, and Isfahan University of Technology (Isfahan, Iran).

This book is divided into two major parts: systems analysis and chemical technology catalog. The five chapters that compose the first part detail the use of linear programming in analyzing the performance of the industry as a whole. The second part of the book is a data base on modern technology for the production of primary feedstocks and intermediate chemicals, plastics and resins, man-made fibers, synthetic rubbers, and thermoplastic elastomers.

The data compiled in this book are widely dispersed in the public and private literature. Exxon Chemical Company assisted in assembling and assessing the market, economic, and technical data.

DALE F. RUDD

Shanahan Valley, Wisconsin
February 1981

Contents

PETROCHEMICAL TECHNOLOGY ASSESSMENT

PART ONE

System Analysis of the United States Industry

CHAPTER ONE

The Chemical Industrial System

INTRODUCTION

The chemical industry is a large, complex, and constantly changing industrial system. There are more than 7000 different organic compounds in commercial production, the majority of which are derived from basic chemicals formed in the processing of petroleum, natural gas, and coal. Before World War II the industry was tied almost completely to steel mills and coke ovens for hydrocarbon raw materials. Now the industry is dependent on petroleum and natural gas as sources of hydrocarbons. For the future, coal and biomass are thought to be important sources.

There are more than 300 U.S. producers of basic and intermediate chemicals. Benzene, toluene, xylene, ethylene, propylene, and propane have more than 25 producers each. Butadiene, styrene, ethylene glycol, paraxylene, acetone, cumene, and phenol have between 12 and 19 producers each. There are three producers of isopropanol, four each of dimethyl terephthalate and caprolactam, five of acrylonitrile, six of linear alkylates, and seven of toluene diisocynate, ethanol, and carbon tetrachloride.

Petroleum companies seeking to upgrade their hydrocarbon raw materials have integrated forward into the petrochemical industry toward fibers, elastomers, plastics, and other consumer products. Rubber, textile, and steel companies seeing synthetic materials as competition for their traditional markets have integrated backward toward the production of synthetic polymers. Fewer than 40% of the major U.S. chemical producers shown in Table 1.1 are true chemical companies; the majority are producers mainly of petroleum and natural gas; metals and minerals; machinery and fabricated metals; food, starches, and beverages; pharmaceuticals; and highly diversified product lines.

These organizations enter the industry for a wide variety of business motivations: to protect a product line, to obtain an outlet for by-products, to upgrade

Table 1.1 Leading U.S. Chemical Companies (Very few are in the chemicals business only)

Company	Chemical % of Total Sales	Company	Chemical % of Total Sales
Du Pont	82	Esmark	6
Union Carbide	57	Gulf Oil	3
Monsanto	80	3M	9
Dow Chemical	69	Pennwalt	45
Exxon	6	Airco	39
Celanese	76	Pfizer	17
Allied Chemical	66	Tenneco	5
W.R. Grace	36	Dow Badische	100
Hercules	83	Kerr-McGee	25
Occidental Petroleum	28	Nalco Chemical	89
Eastman Kodak	20	American Hoechst	65
American Cyanamid	50	El Paso Natural Gas	15
Shell Oil	15	Baychem	63
FMC Corp.	39	Union Oil (Cal.)	7
Phillips Petroleum	19	Uniroyal	8
PPG Industries	33	Cabot	56
Stauffer Chemical	85	GAF Corp.	20
Rohm and Haas	70	Atlantic Richfield	4
NL Industries	41	Borg-Warner	11
Standard Oil (Ind.)	7	Chemetron	47
Akzona	71	General Electric	1
Texaco	4	National Starch	79
Diamond Shamrock	60	Witco Chemical	47
Ethyl	55	Freeport Minerals	88
Cities Service	18	IFF	97
Olin	31	CF Industries	71
Standard Oil (Cal.)	5	Morton-Norwich	30
Goodyear	7	National Distillers	10
B.F. Goodrich	18	Northern Natural Gas	19
Mobil Oil	3	Williams Companies	21
Air Products	74	Arco Polymers	92
Firestone	9	Farmland Industries	13
BASF Wyandotte	71	Eli Lilly	14
Merck	23	Procter & Gamble	2
United States Steel	4	Dart Industries	12
Ashland Oil	12	Dow Corning	78
ICI America	86	Kewanee Oil	66
Lubrizol	99	Texasgulf	34
Continental Oil	6	Houston Natural Gas	33
Reichhold Chemicals	96	Commercial Solvents	90
CIBA-GEIGY	41	Emery Industries	90
IMC	37	Sun Oil	5
Borden	9	Kaiser Aluminum	14

raw materials, to protect sources of materials of manufacture, to exploit special technical and marketing know-how, and simply to enter a profitable line of manufacturing. There is, however, a common feature in this diversity and that is the *technical structure* of the industry.

This book has the objective of defining the inherent technical structure within which the world-wide chemical industry must function. The structure is formed by the large but limited number of chemical syntheses that are available on the commercial scale and by the rigid feedstock, by-product, and energy requirements of these chemical syntheses. The products of one segment of the industry become the feedstocks for other segments of the industry, thereby defining a network of material and energy flows that constrains business activities. Much can be said about the chemical industry from an understanding of this technical structure.

The remainder of this first chapter serves to identify qualitatively the extent of the industry and to define the underlying technical structure. Then in the remaining chapters we develop in detail principles of technology assessment that give insight into the following questions:

1 What are the ultimate performance characteristics of the industry in terms of feedstock, energy, and investment capital use?

2 Is it possible to anticipate the effects on the industry of changing product lines and raw material sources?

3 What are the characteristics of new processes and products that lead to their eventual adoption by the economy?

4 How is the pricing structure of the industry affected by the forces of the market place?

CHEMICALS IN THE ECONOMY

Table 1.2 shows the sales of a reference chemical industry, the major portions of which are organic chemicals derived from petroleum, natural gas, and coal. This table is used to describe the outlet of these chemicals in the economy. We defer discussion of the basic and intermediate chemicals to a later section and focus attention on the main consumer outlets, polymers and additives, agricultural chemicals, fine chemicals, colors, and miscellaneous functional chemicals.

Plastic Materials Plastic materials and synthetic resins are long chain polymers or macromolecules formed by the polymerization of chemical intermediates. They are built up by the chemical combination of low-molecular-weight units termed "monomers" such as ethylene, propylene, styrene, vinyl chloride, and acrylic compounds.

The term *plastic* refers to a relatively tough resin of molecular weight 10,000 to 1,000,000 that can be formed into solid shapes, *resin* refers to polymers used

Table 1.2 Distribution of Products of a U.S. Chemical
Industry (SVA Estimates)

Basic and intermediates	37%
Organics	
Inorganics	
Industrial gases	
Gum and wood chemicals	
Fatty acids	
Polymers and additives	30%
Plastics materials	
Man-made fibers	
Elastomers	
Plasticizers	
Carbon black	
Polymer additives, misc.	
Rubber-processing chemicals	
Agricultural chemicals	10%
Fertilizers	
Pesticides	
Fine chemicals	4%
Medicinals	
Flavors and fragrances	
Colors	3%
Inorganic pigments	
Dyes	
Organic pigments	
Miscellaneous functional chemicals	16%
Adhesives	
Industrial and institutional cleaners	
Surfactants	
Explosives	
Catalysts	
Other	

in coating, adhesives, and binding applications, and *latex* refers to water
dispersions of resins.

The growth of plastics has been through the replacement of natural
materials: wood, glass, paper, and metals. Unique physical and mechanical
properties can be tailored into plastics by variations in chemical structure,
molecular weight, degree of straight chain and branched structure, and the use
of additives.

Table 1.3 shows the end use of plastic materials. We see that the major outlet
is in packaging: flexible bags and wraps, blow-molded bottles, containers,

Table 1.3 Consumption of Major Plastics Materials and Synthetic Resins by End Use (SVA Estimates)

	30% Packaging	23% Construction	11% Housewares	8% Electric and electronic	8% Transportation	8% Paints	4% Toys	4% Furniture	3% Appliances
30% Polyethylene									
Low-density	x	x	x	x	x	x	x	x	x
High-density	x	x	x	x	x	x			
20% Vinyl									
Chloride	x	x	x	x	x	x	x	x	x
Acetate	x					x			
Other	x	x	x		x	x		x	
18% Styrene									
ABS and SAN		x		x	x		x	x	x
Other	x	x	x	x	x		x	x	x
Phenolics	x	x	x	x	x	x	x		x
Polypropylene	x	x	x	x	x		x	x	x
Polyesters		x		x	x	x		x	x
Amino	x	x	x	x		x		x	
Acrylics		x			x	x		x	x
Alkyds						x			
Cellophane	x								
Coumarone indene	x	x					x		
Epoxies		x		x		x			
Cellulosics	x	x		x	x	x			x
Rosin modifications						x			

7

closures, and collapsible tubes. The second outlet is in construction products: panels and siding, flooring, pipe, fittings, conduit, and plywood. Electrical and electronic applications, insulation of wire and cable, is a third major market. The other markets are surface coatings, housewares, automobiles, toys, furniture, and appliances.

Man-Made Fibers The two classes of man-made fiber are the *cellulosics* (rayon and acetates) and the *synthetics* (polyester, nylon, acrylics, and polypropylene). The cellulosics, which are obtained by chemically altering regenerated cellulose from wood pulp and cotton linters, are not classified as petrochemicals. The synthetic fibers are polymerized petrochemical molecules. Generally, the man-made fibers can be tailored for improved washability, durability, strength, and resistance to shrinking and soiling. This has allowed them to replace cotton, wool, and silk in many applications.

Tables 1.4 and 1.5 show the production of man-made fibers and their uses in the economy. Well over half of all fibers consumed are man-made.

Elastomers Synthetic rubbers are polymers that are tailored to be capable of quick recovery from large deformations. Until 1960 there were no synthetic elastomers that duplicated the properties of natural rubber, but now there are polymers available that are equivalent to and superior to natural rubber. Styrene-butadiene rubber, SBR, has wear characteristics superior to natural rubber, is currently less expensive, and is regarded as the general purpose synthetic elastomer. Other synthetics, each with specific properties, include stereospecific polymers, neoprene, butyl, ethylene-propylene-diene monomer (EPDM) terpolymers, nitriles, urethanes, and silicones.

Tables 1.6 and 1.7 show the production and outlet of these synthetic materials. Automotive tires are the major outlet; 65 to 70% of the SBR is used there.

Plasticizers There are about 500 chemicals that can be incorporated into polymers to improve processability or to modify substantially the final product. For example, plasticizers change polyvinyl chloride, a brittle, unworkable polymer, into a highly flexible and workable resin. Most plasticizers are esters

Table 1.4 Man-Made Fibers
(SVA Estimates)

Synthetic	Cellulosic
Polyester, 34%	Rayon, 14%
Nylon, 30%	Acetate, 6%
Acrylic, 9%	
Polyolefins, 6%	

Table 1.5 Consumption of Man-Made Fibers by End Use (SVA Estimates)

35%   Apparel
        Women's dresses
        Men's suits, slacks, and coats
        Women's suits, slacks, and coats
        Shirts
        Women's under- and nightwear
        Apparel linings
        Women's blouses
        Uniforms and work clothes
        Sweaters
        Anklets and socks
        Hosiery
        Robes and loungewear
        Swimwear and other recreational wear
        Men's under- and nightwear
        Other

32%   Home furnishings
        Carpets and rugs
        Draperies and upholstery
        Blankets
        Curtains
        Bedspreads and quilts
        Towels
        Other

11%   Other consumer goods
        Retail piece goods
        Diapers, sanitary napkins, bandages, and other medical
        Other

22%   Industrial tires
        Reinforced plastics
        Rope, cordage, and tape
        Coated fabrics
        Sewing thread
        Other

of carboxylic or phosphoric acids. Esters of phthalic anhydride are the largest group. Tables 1.8 and 1.9 show the production and polymer types into which the plasticizers are added.

Carbon Black Carbon black is a fluffy, finely divided powder consisting of 90–99% elemental carbon. Its major application is as a reinforcing agent in rubber products and minor uses are as pigment in printing ink, plastics, and

Table 1.6 Synthetic Elastomers
(SVA Estimates)

SBR	54%
Stereospecific	24%
Cis-polybutadiene	
Cis-polyisoprene	
EPDM	
Neoprene	7%
Butyl	6%
Nitrile	3%
Urethane	1%
Silicone	1%
Other	

Table 1.7 Consumption of Natural and
Synthetic Rubber (SVA Estimates)

Tires and related	66%
Molded goods	
Automotive	5%
Other	5%
Foam rubber	3%
Shoe products	2%
Hose, tubing	2%
Rubber footwear	2%
O-rings, packing gaskets	2%
Sponge rubber products	2%
Solvent and latex cement	1%
Belts, belting	1%
Wire, cable	1%
Coated fabrics	1%
Floor and wall coverings	1%
Pressure-sensitive tapes	
Industrial rolls	
Athletic goods	
Military goods	
Thread (bare)	
Drugs, medical sundries	
Toys, balloons	
Other	

Table 1.8 Production of Plasticizers
(SVA Estimates)

Phthalates	67%
Dioctyl (DOP)	
Diisodecyl (DIDP)	
Diisooctyl (DIOP)	
Dibutyl	
Ditridecyl	
Diethyl	
n-Hexyl, *n*-decyl	
Butyl octyl	
Dimethyl	
Other	
Phosphates	7%
Tricresyl	
Cresyl diphenyl	
Acyclic	
Other cyclic	
Epoxies	7%
Epoxidized soya oils	
Octyl epoxy tallates	
Low-temperature	5%
Adipates	
Azelates	
Sebacates	
Polymeric	3%
Other	
Stearates	
Oleates	
Trimellitates	
Other acyclic	
Other cyclic	

paints. This is obtained by the partial combustion and thermal decomposition of hydrocarbon liquids and gases. Table 1.10 shows the end uses.

Polymer Additives Flame retardants, heat stabilizers, lubricants, antioxidants, organic peroxides, radiation absorbers, and antistatic agents are added to polymers in addition to plasticizers. Many of the materials shown in Table 1.11 are of petrochemical origin.

Table 1.9 Consumption of Plasticizers by End Use (SVA Estimates)

Polyvinyl chloride	64%
Film and sheet	
Molding and extrusion	
Wire and cable	
Flooring	
Paper and textile coatings	
Other	
Other plastics	10%
Exports	10%
Nonplastics uses	16%

Table 1.10 Consumption of Carbon Black (SVA Estimates)

Rubber	94%
Printing ink	3%
Plastics	
Paint	
Paper	
Other	

Rubber-Processing Chemicals Raw rubber has few applications. Antioxidants, stabilizers, vulcanizing agents, and the other chemicals shown in Table 1.12 must be used to convert raw rubber into a useful product.

Fertilizers Of the three primary nutrients required for plant growth, nitrogen, phosphorus, and potassium, nitrogen is of petrochemical origin. It is supplied in the form of ammonia, ammonium nitrate, ammonium sulfate, ammonium phosphates, or urea. These compounds trace their origins to hydrocarbons from which ammonia is synthesized in enormous amounts.

Pesticides The three major types of pesticide are (1) herbicides, (2) insecticides, and (3) fungicides. Synthetic organic compounds play a major role here, although inorganic and botanical products are important. Table 1.13 shows the production and use of major pesticides.

Medicinals A large number of the synthetic organic medicinals shown in Table 1.14 are chemical products. Synthetic organic and fermentation products

Table 1.11 Consumption of
Miscellaneous Polymer Additives
(SVA Estimates)

Flame retardants	50%
Additives	
Phosphate esters	
Nonhalogenated	
Halogenated	
Chlorinated paraffins	
Antimony oxide	
Bromine compounds	
Boron compounds	
Other	
Reactive intermediates	
Urethane	
Polyester	
Epoxy	
Other	
Heat stabilizers	20%
Barium-cadmium	
Lead	
Organotins	
Calcium-zinc	
Lubricants	15%
Metallic stearates	
Fatty acid amides	
Fatty acids and esters	
Waxes	
Antioxidants	6%
Organic peroxides	6%
Ultraviolet radiation absorbers	
Antistats	

amount to 48% of the production. All the antibiotics are partially or fully produced by fermentation, except chloramphenicol and cycloserine. Most vitamins are made by chemical synthesis, except B_{12}. In the central depressants and stimulants category, aspirin (acetylsalicylic acid) is the most widely used medicinal. Choline chloride is used principally as a dietary supplement for chicken and turkey feeds.

Flavors and Fragrances Aroma chemicals, flavor enhancers, and sweeteners are the three classifications of flavors and fragrances of synthetic origin shown

Table 1.12 Production of Synthetic Organic Rubber-Processing
Chemicals (SVA Estimates)

Antioxidants, antiozonants, and stabilizers	54%
Amino compounds	
Substituted *p*-phenylenediamines	
n-Phenylamine	
Octyl diphenylamine	
Other	
Phenols and phenol-phosphites	
Alkylated phenol	
Polyphenolics and bisphenols	
Styrenated phenol	
Other	
Accelerators, activators, and vulcanizing agents	35%
Thiazole derivates	
2,2′-Dithiobis (benzothiazole) (MBTS)	
2-Mercaptobenzothiazole (MBT)	
n-Cyclohexyl-2-benzothiazole sulfenamide	
2-Mercaptobenzothiazole, zinc salt	
Other	
Thiruams	
Dithiocarbomates	
Aldehyde-amine reaction products	
Other	
Other	11%
Dodecyl mercaptan	
Sodium-dimethyldithiocarbamate	
Other	

---

in Table 1.15. Many of these are not petrochemicals, such as monosodium
glutamate (MSG). The majority of fragrance compounds are synthetic.

Dyes and Organic Pigments   The textile industry is the largest consumer of
synthetic organic dyes. Twenty percent is used by the paper industry and lesser
amounts are used in the manufacture of plastics, food, gasoline, and organic
pigments. The major outlets of pigments are printers ink, paints, and plastics.
The production is seen in Table 1.16.

Adhesives   Synthetic adhesives consist of thermoplastic and thermosetting
resins and elastomers. Phenolics, amino resins, epoxies, polyvinyl acetate, and
synthetic hot melt adhesives are significant petrochemical products. The synthetic adhesives may be purchased directly or formulated "captively" by the
user. For example, a manufacturer of particle board may purchase the makings
for a phenolic, a urea and melamine formaldehyde, or a polyvinyl acetate

Table 1.13 Production and Uses of Major Pesticides (SVA Estimates)

Pesticide		Major Uses
Herbicides	37%	
Atrazine (AAtrex)		Pre-emergence corn and sorghum
2,4-D		Corn and wheat
Methanearsonates (DSMA, MSMA)		Post-emergence cotton
Propachlor, CDAA (Ramrod, Randox)		Pre-emergence corn and rice
Trifluralin (Treflan)		Pre-emergence cotton and soybean
Amiben		Pre-emergence soybean
Alachlor (Lasso)		Pre-emergence corn and soybean
Bromacil (Hyvar, Krovar)		Citrus and pineapple
Dicamba (Banvel)		Small grains
Diuron		Pre-emergency corn and cotton
Propanil		Rice
Sutan		Pre-emergence corn
2,4,5-T		Brush control
Insecticides	51%	
Chlorinated hydrocarbons		
Paradichlorobenzene		Fumigant
Toxaphene		Cotton foliage
DDT		Broad spectrum, especially cotton, livestock, & grain
Chlordane		Household and institutional
Aldrin		Broad spectrum, especially soil
Methoxychlor		Broad spectrum, especially cattle
Heptachlor		Broad spectrum, especially cotton
Organophosphorus		
Methyl parathion		Cotton foliage
Malathion		Broad spectrum
Parathion		Broad spectrum
Diazinon		Corn soil
Disulfoton (Di-Syston)		Systemic soil, especially cotton and potato
Phorate (Thimet)		Systemic soil, especially cotton and corn
Carbamates		
Carbaryl (Sevin)		Broad spectrum
Carbonfuran (Furadan)		Corn
Bux Ten		Corn soil
Brominates and other		
Methyl bromide		Fumigant
Warfarin		Rodents
DBCP		Nematodes, fumigant
Fungicides	12%	
Pentachlorophenol and salts		Wood preservative
Captan		Apple foliage
Dithiocarbamates		Potato and tomato foliage

Table 1.14 Synthetic Organic Bulk Medicinals (SVA Estimates)

Antibiotics	7%
Penicillins	
Tetracyclines	
Other	
Vitamins	13%
B complex	
E	
C	
A	
Other	
Anti-infective agents	15%
Antiprotozoan agents	
Anthelmintics	
Sulfonamides	
Urinary antiseptics	
Other	
Central depressants and stimulants	22%
Analgesics and antipyretics	
Barbiturates	
Other	
Gastrointestinal agents and therapeutic nutrients	34%
Choline chloride	
Amino acids and salts	
Other	
Hormones and substitutes	
Autonomic drugs	
Sympathomimetic	
Other	
Antihistamines	
Dermatological agents and local anesthetics	
Salicylic acid	
Other	
Expectorants and mucolytic agents	
Renal-acting and edema-reducing agents	

Table 1.15 Synthetic Flavor and
Perfume Materials (SVA Estimates)

Aroma chemicals	50%
Benzyl alcohol	
Methyl salicylate	
Terpineols	
Anethole	
Cinnamaldehyde	
Benzyl acetate	
Geraniol	
p-Anisaldehyde	
Citronellol	
Isobornyl acetate	
α-Terpinyl acetate	
Menthol	
Methylionones	
Ethyl butyrate	
Eugenol	
Other	
Flavor enhancers	45%
MSG	
Sweeteners	5%
Saccharin	
Cyclamates	

adhesive and manufacture the adhesive for inplant use. Table 1.17 represents the shipment of adhesives by type.

Cleaners and Surfactants Detergents contain surface-active compounds such as the alkylbenzene sulfonates, the linear alcohols, and carboxylic acid salts which may be of largely petrochemical origin. The production is seen in Table 1.18.

Other Functional Specialties Other outlets of petrochemical compounds are listed in Table 1.19.

Technology Assessment

The thousands of different chemical formulations that enter the consumer market are formed from a smaller but still large number of chemical building blocks. These chemical building blocks in turn are synthesized from intermediate chemicals which are derived from simple primary feedstocks. The manufacture of these primary, intermediate, and end chemicals is accom-

Table 1.16 Organic Pigments
(SVA Estimates)

Toners	95%
Red	
Lithol Red R	
Permanent Red 2B	
Lake Red C	
Toluidine	
Eosin	
Lithol Red 2G	
Lithol Rubine B	
Naphthol reds	
Other	
Blue	
Phthalocyanine	
Alkali	
Other	
Yellow	
Benzidines	
Hansa	
Other	
Green	
Phthalocyanine	
Other	
Violet	
Orange	
Black	
Brown	
Lakes	5%
Red	
Violet	
Other	

plished through several hundred well-defined chemical processes. In this section we focus attention on these chemicals and processes that define the intermediate part of the chemical industry. Figure 1.1 illustrates *part* of the technology linking primary feedstocks to consumer goods.

Figure 1.2 illustrates a fundamental and common problem in process technology assessment that arises when the technology performs similar but not identical tasks. In this example four processes for the manufacture of acetic acid, an intermediate chemical used mainly in the manufacture of cellulose acetate and vinyl acetate, interact with the surrounding industry in drastically different ways. There is not a raw material common among the four alternatives and different by-products appear. An accurate assessment of these

Table 1.17   Adhesives (SVA Estimates)

Synthetic	50%
From purchased resins	
From captively produced resins	
Rubber cement	12%
Sizes	13%
Rosin	
Other	
Other	25%
Animal glues	
Starches and dextrins	
Natural gums and resins	
Casein	
Other	

alternatives can only be made by determining the influence of these dissimilar interactions with the industry as a whole.

In the petrochemical industry it is common to find dissimilar chemical routes to the wide variety of products. Table 1.20 shows the number of commercial processes involving different raw materials and/or different co-products. These are just a few of the hundreds of examples occurring in this book.

Clearly the small manufacturer need only estimate the prices of the raw materials, energy, and investment, and the value of the products to make a reasonable economic assessment of the alternatives. The small manufacturer will not sufficiently disrupt the supply and demand balances in the industry to make this simple economic analysis unreliable.

However, when decisions are to be made that will have a significant impact on the industry, a much more sophisticated method of analysis is required. In such a case great disruptions can occur that would drastically change the future structure of the industry. Somehow, these structural changes must be assessed accurately and must play an important role in economic and technology assessment.

Figure 1.3 shows our perception of the petrochemical industry. Feedstocks enter the industry from a limited exogenous supply, largely as by-products from the energy industries. These feedstocks enter a network of chemical processes that constitute the petrochemical industry and are converted into intermediate and final chemicals. These chemicals then meet the exogenous demand for fiber, plastics, elastomers, and hundreds of other petrochemical products.

Several models of industrial performance have proved to be useful. All models are based on the following assumptions:

Table 1.18 Surfactants (SVA Estimates)

Anionics   68%
  Sulfonates
    Dodecylbenzene sulfonates
    Other alkylbenzene sulfonates
    Lignosulfonates
    Benzene-, cumene-, toluene-, and xylenesulfonates
    Naphthalenesulfonates
    Other

  Carboxylic acid salts
    Amine salts and salts with amide, ester, or ether linkages
    Tallow
    Coconut
    Other

  Sulfates
    Ethers
    Alcohols
    Natural fats and oils
    Acids, amides, esters

  Phosphates and polyphosphates

  Other anionic

Nonionics   26%
  Ethylene oxide condensates
    Linear alcohol
    Alkylphenol
    Anhydrosorbitol esters
    Natural fats and oils
    Tridecyl alcohol ether

  Carboxylic ester
    Glycerol
    Polyethylene glycol
    Anhydrosorbitol
    Other

  Fatty acid alkanolamides

  Other nonionics

Cationics   6%
  Quaternary ammonium salts
  Other cationics

Amphoterics

1 The chemical industry as the whole can be visualized and represented as a network of chemical processes.

2 The processes involve chemical and/or physical transformation and are therefore defined by their specific material flows. Each process can be thought of as an entry in a technology catalog (see Part 2 of this book).

3 The material interactions among processes are linear. The processing network therefore can be represented as a linear input/output matrix.

4 There exists a specific supply/demand environment and process capacity limitations.

In the models, the overall industry is assumed to seek to utilize its available resources in an optimal fashion. The models differ in their perception of optimality.

Table 1.19 Miscellaneous Functional Chemical Specialties (SVA Estimates)

Multipurpose functional specialties
    Thickeners
    Fluorocarbons
    Antioxidants
    Biocides
    Corrosion inhibitors
    Flame retardants
    Enzymes
    Metallic soaps
    Chelating agents
    Ultraviolet radiation absorbers

Market-directed specialties
    Petroleum additives
    Water-management chemicals
    Automotive chemicals
    Metal-finishing chemicals
    Diagnostic aids
    Food additives
    Photographic chemical preparations
    Laboratory chemicals
    Paper chemicals
    Textile chemicals
    Cosmetic specialties
    Paint additives
    Drilling mud additives
    Foundry chemicals
    Flotation reagents
    Printing chemicals

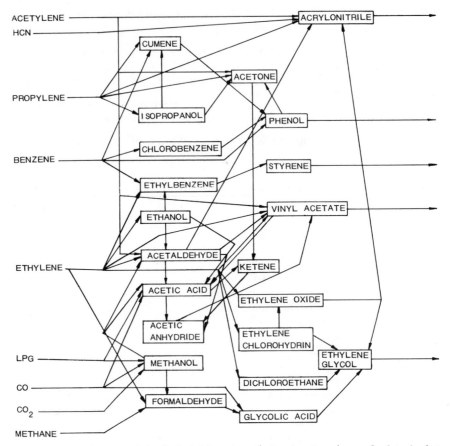

Figure 1.1 Only part of the industrial system that converts primary feedstocks into end chemicals.

ECONOMIC MODEL I

The economic model of the U.S. industry includes 182 processes involved in the transformation of 131 chemical intermediates and feedstocks. Within the specified process capacity limits B_j and feedstock supply constraints S_i, the objective function basically seeks out processes to operate at levels X_j, such that the product demands D_i are satisfied at a minimum total production cost to the industry.

The production cost to the industry is defined as

$$\text{production cost} = \sum_{i=1}^{N} P_i F_i + \sum_{j=1}^{M} C_j X_j + \sum_{i=1}^{N} (Q_i - D_i)(P_i - H_i)$$

where C_j is the total cost of production for process j, and P_i and H_i are chemical prices and heating values for chemical i, respectively. The term C_j is composed of raw material costs M_j, utilities cost U_j, labor-related costs L_j, and an investment-related cost I_j. By-products are valued as credits and are included in the raw material cost. A return on investment element is included in the investment-related cost. The term P_iF_i represents cost of feedstock materials,

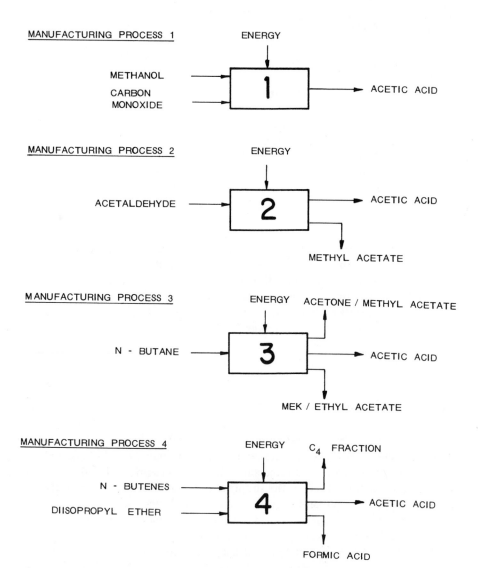

Figure 1.2 The fundamental problem in technology assessment. To produce the same product, these four processes interact with the industry as a whole in drastically dissimilar ways.

Table 1.20 Examples of Products That Can Be
Manufactured by Processes Involving Different
Raw Materials and/or By-Products

Product	Number of Commercial Processes with Different Raw Materials and Products
Acetaldehyde	8
Acetic acid	15
Acetic anhydride	3
Acetone	10
.	.
.	.
.	.
Vinyl acetate	3
Vinyl chloride	3
m-Xylene	3
o-Xylene	1
p-Xylene	2

and the term $(Q_i - D_i)(P_i - H_i)$ represents a correction term for surplus chemicals; that is, any surplus chemicals $Q_i - D_i$ resulting from the model network will only have heating value instead of chemical value. Table 1.21 is an example of process economic data for the phenol via air oxidation of cumene process. Chemical prices P_i and heating values H_i are estimated for all the products for 1975 and 1977. To use the model for 1985, or other years, new cost data need to be estimated or, alternatively, these cost data could be generated by using general escalation factors.

INTEGRATED MODEL II

In the U.S. industry, market prices of intermediate chemicals are a major deciding factor in process selection and Model I accounts for this explicitly by including price estimates and projections. It is well recognized that the prices of intermediate chemicals reflect the supply and demand for those chemicals and that the supply and demand reflect the existing industry-wide capacity. Model II eliminates the use of price projections for intermediate chemicals by assuming the model industry is completely integrated.

The integrated total production cost is formulated as follows:

$$\text{total production cost} = \sum_{i=1}^{N} P_i F_i + \sum_{j=1}^{M} C_j^* X_j - \sum_{i=1}^{N} H_i(Q_i - D_i)$$

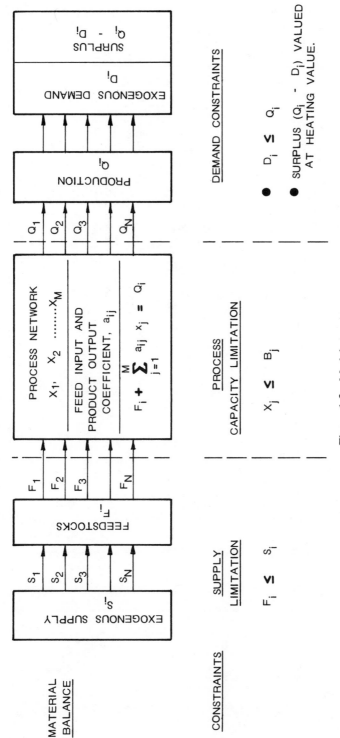

Figure 1.3 Model structure.

Table 1.21 Production Cost of Phenol Via Air Oxidation of Cumene (Typical Plant Size 200 Million lb/yr)

I. Input/Output Data—Feedstocks, Products, and By-Products (lb/lb of Phenol)

1. Phenol	1.00	5. Hydrogen	-0.001
2. Cumene	-1.35	6. Sodium carbonate	-0.003
3. Caustic soda	-0.0099	7. Carbon dioxide	-0.002
4. Sulfuric acid	-0.01	8. Acetone	0.61

II. Process Cost Data (1977 Basis)

A. Feedstocks and Chemicals Costs Less By-Product Credits

	¢/lb	¢/lb of Phenol		¢/lb	¢/lb of Phenol
1. Cumene	13	17.55	5. Sodium Carbonate	4.8	0.015
2. Caustic soda	7.5	0.074	6. Carbon dioxide	0.30	0.000
3. Sulfuric acid	2.25	0.023	7. Acetone	14.5	-8.845
4. Hydrogen	25	0.025	8. Fuel		-0.09

Total = 8.75 ¢/lb of phenol

B. Utilities Costs

	Requirement¹	Unit Cost, ¢	¢/lb of Phenol		Requirement¹	Unit Cost, ¢	¢/lb of Phenol
1. Electricity, KWH	0.139	1.7	0.236	4. Water-Process/Cooling			0.057
2. Steam, lbs	5.05	0.22	1.111	5. Inert gas, SCF	0.04	0.04	0.0016
3. Fuel, MBTU	0.0002	200	0.04				

Total = 1.45 ¢/lb of phenol

C. Fixed Costs

1. Labor-related costs² = 1.42		¢/lb of phenol
2. Investment-related costs³ = 9.89		¢/lb of phenol

Total production cost (A + B + C) = 21.51 ¢/lb of phenol

¹Requirement = consumption per pound of phenol.
²Labor-related costs = Operating labor, maintenance labor, control laboratory, and plant overhead.
³Investment-related costs = ROI, maintenance materials, operating supplies, taxes, insurance, and depreciation.

26

The terminology is the same as in Model I, except C_j^* represents utilities, investment, and labor-related costs as well as the fixed costs associated with the raw materials portion of the process such as catalysts, additives, and fillers. Because no intermediate prices are assigned and therefore no by-product credits are given, the surplus materials are credited only at heating value. This model gives an interesting view of the limits of performance of a completely integrated industry.

CONCLUDING REMARKS

We conclude this introductory chapter of Part 1 with a synopsis of the studies to be discussed in subsequent chapters, as well as some comments on Part 2.

Chapter 2   Economic Model of the Intermediate U.S. Chemical Industry

Model I is applied to a projected 1985 U.S. industry consisting of 131 chemicals and 182 processes. We show how the model reproduces the known 1977 industry, what changes ought to occur to meet the needs of a 1985 industry, how new chemical technology might be assessed for the 1985 industry, what effect perturbations in energy consumption might have, and how perceived shortages in primary feedstocks should be assessed.

Chapter 3   Integrated Model of the Intermediate U.S. Chemical Industry

We repeat many of the studies of a 1985 industry performed in Chapter 2, and show the effect of changes in the long-range objective on the technology assessment. We also include a study of the subsidy required to make fermentation alcohol a competitive feedstock for chemicals manufacture.

Chapter 4   Systems Study of Interchangeable Chemical Products

The previous chapters focused on the intermediate chemicals industry. The study is expanded to include the functions performed by chemicals as consumer products. Added are 117 new processes for the manufacture of 72 polymer end products. The end products include groups of polymers such as epoxy resins, barrier resins, fibers, elastomers, and other product groups within which wide substitutions are possible. We show how insight into future product trends is accessible through product technology assessment.

Chapter 5   Pricing Structure of the U.S. Chemical Industry

It is shown that the *shadow prices* of an integrated model are closely related to the market prices for chemicals. We show how the forces of the market place tend to adjust the prices of chemicals. This is also exemplified by a study on the impact of restrictions on toxic chemicals on the production of synthetic materials.

Part 2 Chemical Technology Catalog

Presented in the catalog is an enormous accumulation of data on hundreds of chemical processes. The data base is used in all of the studies summarized in the early parts of this book.

REFERENCES

Related Works

The work presented in this book is a part of the extensive studies of the petrochemical industry system found in the following references.

S. Fathi-Afshar, "A Systems Study of the Interchangeable Petrochemical Products," PhD dissertation, University of Wisconsin-Madison, 1979.

S. Fathi-Afshar, D. S. Maisel, D. F. Rudd, A. A. Trevino, and W. W. Yuan, "Advances in Petrochemical Technology Assessment," to appear in *Chemical Engineering Science,* 1981.

S. Fathi-Afshar and D. F. Rudd, "Biomass Ethanol as a Chemical Feedstock in the United States," *Biotechnology and Bioengineering,* **22,** 677–679 (1980).

S. Fathi-Afshar and D. F. Rudd, "Technology Assessment of the Polymer Products Industry," *Polymer-Plastics Technology and Engineering,* **14**(2), 135–160 (1980).

S. Fathi-Afshar and D. F. Rudd, "The Economic Impact of New Chemical Technology," *Chemical Engineering Science*, 1981.

J. K. Mikkelsen, "A System Study of the Oil and Petrochemical Industry in Western Europe with Special Considerations to this Future Industry in Norway," Master's thesis, University of Wisconsin-Madison, 1979.

J. K. Mikkelsen and D. F. Rudd, "Development of a Norwegian Petrochemical Industry," *Engineering and Process Economics,* 1981.

D. F. Rudd, "Modelling the Development of the Intermediate Chemicals Industry," *Chemical Engineering Journal,* **9** (1975), also in *Adaptive Economic Models*, R. H. Day, Ed., Academic Press, New York, 1975.

B. O. Palsson, S. Fathi-Afshar, D. F. Rudd and E. N. Lightfoot, "Economic Evaluation of the Large-Scale Use of Biomass as a Source of Feedstocks for the U.S. Chemical Industry," *Science,* 1981.

R. N. Saxena, "An Approach to the Simulation of the Development of U.S. Petrochemical Industry," PhD dissertation, University of Wisconsin-Madison, 1978.

A. Sophos, E. Rotstein, and G. Stephanopoulos, "Thermodynamic Bounds and the Selection of Technologies in the Petrochemical Industry," *Chemical Engineering Science,* **35,** 1049–1065 (1980).

A. Sophos, E. Rotstein, and G. Stephanopoulos, "Multiobjective Analysis in Modeling the Petrochemical Industry," *Chemical Engineering Science*, 1980.

M. A. Stadtherr, "A Systems Study of the Petrochemical Industry," PhD dissertation, University of Wisconsin-Madison, 1976.

M. A. Stadtherr, "A Systems Approach to Assessing New Petrochemical Technology," *Chemical Engineering Science*, **33**, (1978).

M. A. Stadtherr and D. F. Rudd, "Systems Study of the Petrochemical Industry," *Chemical Engineering Science*, **31**, (1976).

M. A. Stadtherr and D. F. Rudd, "Resource Management in the Petrochemical Industry," *Management Science*, **24**(7), (March 1978).

M. A. Stadtherr and D. F. Rudd, "Resource Use by the Petrochemical Industry," *Chemical Engineering Science*, **33**, (1978).

A. A. Trevino, "Integrated Systems Study of the Development of the Mexican Petrochemical Industry," PhD dissertation, University of Wisconsin-Madison, 1979.

A. A. Trevino and D. F. Rudd, "On Planning an Integrated Mexican Petrochemical Industry," *Engineering and Process Economics*, **5**, 129-142 (1980).

The following studies are along the lines similar to the earlier studies of iron and steel industries, coal industry, and petroleum refining industry.

M. A. Abe, "Dynamic Macroeconomic Models of Production, Investment, and Technological Change of the U.S. and Japanese Iron and Steel Industries," PhD dissertation, University of Wisconsin-Madison, 1970.

R. H. Day and J. P. Nelson, "A Class of Dynamic Models for Describing and Projecting Industrial Development," *Journal of Econometrics*, **1**, 155-190 (1973).

T. Fabian, "A Linear Programming Model of Integrated Iron and Steel Production," *Management Science*, **4**(4), (July 1958).

T. Fabian, "Process Analysis of the U.S. Iron and Steel Industry," in *Studies in Process Analysis*, A. S. Manne and H. M. Markowitz, Eds., Wiley, New York, 1963.

J. M. Henderson, "Efficiency and Pricing in the Coal Industry," *Review of Economics and Statistics*, **38**, 50-60 (1956).

M. A. Lindsay, "A Simulation Model of the Development of Petroleum Refining Capacity," Social Systems Research Institute Workshop Series, No. 7412, University of Wisconsin-Madison, August 1974.

A. S. Manne, "A Linear Programming Model of the U.S. Petroleum Refining Industry," in *Studies in Process Analysis*, A. S. Manne and H. M. Markowitz, Eds., Wiley, New York, 1963.

W. K. Tabb, "A Recursive Programming Model of Resource Allocation and Technological Change in the U.S. Bituminous Coal Industry," PhD dissertation, University of Wisconsin-Madison, 1968.

C. S. Tsao and R. H. Day, "A Process Analysis Model of the U.S. Steel Industry," *Management Science*, **17**(10), (June 1971).

Technical and Marketing Reference Works

The following are some selected reference works concerning the chemical process industries.

American Chemical Society, *Chemistry in the Economy*, Washington, D.C., 1974. An excellent book covering the social and economic impacts of chemistry, accomplishments of chemistry, and its future developments in different sectors of the economy.

N. M. Bikales, Ed., *Encyclopedia of Polymer Science and Technology* (15 volumes, 2 supplement volumes, 1 index volume), Wiley-Interscience, New York, 1964-1977. A comprehensive source of information on chemistry and the manufacture of plastics and resins, rubbers, and fibers.

A. M. Brownstein, *Trends in Petrochemical Technology*, Petroleum Publishing Company, Tulsa, Oklahoma, 1976. Production and derivatives of aromatics and olefins and their dependence on fossil fuels are considered. The book provides a critique of current process economics and describes potential competitive technologies. A good history of production and sales volumes, and processing capacities are reported.

L. W. Codd, K. Dijkhoff, J. H. Fearon, C. J. van Oss, H. G. Roebersen, and E. G. Stanford, Eds., *Chemical Technology: An Encyclopedic Treatment* (8 volumes), Barnes & Noble, Inc., New York, 1968-1975. Also *Materials and Technology*, published by Longman, London and J. H. deBussy, Amsterdam. A good source of information on the technology, uses, and economics of a number of industrial chemicals.

M. Grayson, Ed., *Kirk-Othmer Encyclopedia of Chemical Technology* (8 volumes, 1 index volume), 3rd ed., Wiley-Interscience, New York, 1978-1979. This work is the updated version of the 2nd ed. of the *Kirk-Othmer Encyclopedia of Chemical Technology*, A. Standen, Ed.

A. V. G. Hahn, *The Petrochemical Industry*, McGraw-Hill, New York, 1970. Despite some out-of-date material on economics and market structures, often a useful book in getting familiar with the chemical processing industries.

F. A. Lowenheim and M. K. Moran, *Faith, Keyes, and Clark's Industrial Chemicals*, 4th ed., Wiley-Interscience, New York, 1975. A good summary of common industrial chemicals with a history of sales volume and prices.

H. F. Mark, S. M. Atlas, and E. Cernia, *Man-Made Fibers: Science and Technology* (3 volumes), Wiley-Interscience, New York, 1967-1968. Chemistry and the manufacture of major man-made fibers are discussed.

J. J. McKetta, Ed., *Encyclopedia of Chemical Processing and Design* (11 volumes), Marcel Dekker, New York, 1976-1980. This work provides some useful information on history, manufacturing, and the design aspects of the chemical processing industries.

M. Morton, Ed., *Rubber Technology*, 2nd ed., Van Nostrand Reinhold, New York, 1973. This book discusses historical development, properties, processing, applications, and consumption of various types of rubbers and compounding ingredients.

P. Noble, Ed., *The Kline Guide to the Chemical Industry*, 2nd ed., Charles H. Kline, Fairfield, N.J., 1974. An excellent book, providing a comprehensive analysis of the U.S. chemical industry, its companies, products, economics and profitability, and sources of information.

B. G. Reuben and M. L. Burstall, *The Chemical Economy*, Longman, London, 1973. Although it puts emphasis on the British chemical industry, the book presents a good summary of the growth, economics, and the products of the chemical industry. Despite some out-of-date material, it provides some useful historical data on production, sales, economics, and prices in the United Kingdom as well as the United States.

D. W. Rosato, W. K. Fallon, and D. V. Rosato, *Markets for Plastics*, Van Nostrand Reinhold, New York, 1969. This book reviews existing and potential markets and applications for plastics in the United States and abroad. Despite some out-of-date material, it provides some statistics on production, sales, and consumption pattern of plastics materials.

R. N. Shreve, and J. A. Brink, Jr., *Chemical Process Industries*, 4th ed., McGraw-Hill, New York, 1977. Its good introduction and historical background of the particular process, and consideration of uses and economics, make this book a good summary of the chemical process industries.

M. Sittig, *Organic Chemical Process Encyclopedia*, Noyes Development Corporation, Park Ridge, N.J., 1967. This one-volume book contains 587 flow charts covering the manufacture of every important organic chemical. In many of these flow charts, pertinent data based upon industrial collaboration are presented for plant design, equipment, operating conditions, yields, and economics.

A. Standen, Ed., *Kirk-Othmer Encyclopedia of Chemical Technology*, 22 vols., Wiley-Interscience, New York, 1972. Undoubtedly the most useful work containing articles on virtually every aspect of the chemical industry. The third edition of this work has been edited by M. Grayson.

R. B. Stobaugh, Jr., *Petrochemical Manufacturing & Marketing Guide*, 2 vols., Gulf Publishing Company, Houston, Tex., 1966. Although it is out of date now, it provides a summary of marketing, producing companies, historical data, and economics for aromatics and derivatives, olefins, diolefins, and acetylene.

A. L. Waddams, *Chemicals from Petroleum*, 3rd ed., Halsted Press, New York, 1973. Characteristics of petroleum chemical manufacture are discussed. Besides a summary of the manufacture of some important organic chemicals, it provides some historical data on the production and consumption pattern of chemicals in the United States, United Kingdom, and some Western European countries.

J. Wei, T. W. F. Russell, and M. W. Swartzlander, *The Structure of the Chemical Processing Industries*, McGraw-Hill, New York, 1979. This textbook provides some statistics on the chemical production, consumption, and prices.

H. Wittcoff and B. Reuben, *Industrial Organic Chemicals in Perspective*, Wiley-Interscience, New York, 1980.

Trade and Technical Journals

There are several hundred trade and technical periodicals serving the chemical industry. The following are some selective journals of interest to marketing and management personnel.

Chemical Age, Morgan-Grampian House, London. A weekly journal emphasizing the British industry; it also reports on world-wide news and technical developments of the chemical process industries.

Chemical & Engineering News, American Chemical Society, Washington, D.C. This very important weekly publication reports on the most recent statistics and covers developments in engineering, research, marketing, and production of the entire chemical industry.

Chemical Engineering, McGraw-Hill Publications, Hightstown, N.J. This biweekly publication is particularly useful for information on the new processes.

Chemical Engineering Progress, American Institute of Chemical Engineers, New York. This monthly journal is a good source of information on economics and technical aspects of the new technologies.

Chemical Marketing Reporter (formerly *Oil, Paint and Drug Reporter*), Schnell, New York. This weekly newspaper is the best source of the current market prices of chemicals. It also contains articles on markets, plant facilities, and general developments of the chemical industries.

Chemical Week, McGraw-Hill Publications, Hightstown, N.J. This is one of the major journals published weekly, reporting on business, economics, marketing, technical aspects, and international developments of the chemical industry.

Elastomerics (formerly *Rubber Age*), Communication Channels, Inc., Atlanta, Ga. This monthly journal features technical papers, and reports on markets, statistics, and rubber industry news.

Hydrocarbon Processing, Gulf Publishing Company, Houston, Tex. This monthly published journal is a leading technical and trade magazine of the oil-processing and petrochemical industries. Of particular value is the annual review of the petrochemical industry in the November issue, containing very useful information on the economics, technical aspects, and commercial installations of the processes leading to petrochemical products.

Modern Plastics, McGraw-Hill Publications, Hightstown, N.J. This leading monthly publication contains articles on technical and marketing news. Of particular value is the January issue outlook, presenting valuable statistics on supplies, sales, demand breakdown, producing companies, and prices of plastics materials.

Oil & Gas Journal, Petroleum Publishing Company, Tulsa, Okla. This weekly publication is an important journal reporting on the technology and business news of the oil and gas industry. It contains useful statistics on production and prices.

Plastics World, Cahners Publishing Company, Boston. This monthly journal reports on technical and marketing news of the plastics manufacturing industries.

Rubber World, Bill Communications, Inc., Akron, O. This monthly publication is one of the leading trade journals of the rubber manufacturing industries, covering worldwide technical and marketing news.

Textile Organon, Textile Economics Bureau, Inc., New York. This monthly publication is an excellent source of information, especially important to the U.S. man-made fiber industry. It provides market data on production, sales, and consumption pattern of the man-made fibers.

Private Sources

Several consulting organizations compile information for sale on detailed process economics, processing capacities, supply and demand balances for specific groups of chemicals, historical data and forecasts and possible interactions among various sectors of the industry. Most chemical companies subscribe to these compilations and initiate studies on their own. This information is generally only available through industrial sources.

U.S. Tariff Commission, *Synthetic Organic Chemicals*, U.S. Government Printing Office, Washington, D.C. This annual publication provides detailed reports on production and sales of organic chemicals.

Miscellaneous References

The following references were also used in the course of our studies.

Chemical & Engineering News, **54**(13), 6 (1976).

H. G. Daellenbach and E. J. Bell, *User's Guide to Linear Programming*, Prentice-Hall, Englewood Cliffs, N.J., 1970.

G. B. Dantzig, *Linear Programming and Extensions*, Princeton University Press, Princeton, N.J., 1974.

J. G. da Silva, G. E. Serra, J. R. Moreira, J. C. Concalves, and J. Goldemburg, "Energy Balance for Ethyl Alcohol Production from Crops," *Science,* **201** (September 1978).

N. J. Driebeek, *Applied Linear Programming*, Addison-Wesley Publishing Company, Reading, Mass., 1969.

D. Gale, *The Theory of Linear Economic Models,* McGraw-Hill, New York, 1960.

S. I. Gass, *Linear Programming,* 4th ed., McGraw-Hill, New York, 1975.

R. Greene, "Carbonated-drink packs: some to fizzle out?," *Chemical Engineering,* July 18, 1977.

G. F. Hadley, *Linear Programming*, Addison-Wesley, Reading, Mass., 1962.

L. R. Lindeman and C. Rocchiccioli, "Ethanol in Brazil: Brief Summary of the State of Industry in 1977," *Biotechnology and Bioengineering,* **XXI**, 1107–1119 (1979).

D. S. Maisel, "Trends in the Petrochemical Industry," *Chemical Engineering Progress* (January 1980).

D. A. O'Sullivan, "UN Workshop Urges Wider Use of Ethanol," *Chemical & Engineering News* (April 1979).

R. H. Perry and C. H. Chilton, Eds., *Chemical Engineers' Handbook,* 5th ed., McGraw-Hill, New York, 1973.

R. C. Reid, J. M. Prausnitz, and T. K. Sherwood, *The Properties and Gases and Liquids*, 3rd ed., McGraw-Hill, New York, 1977.

M. Sittig, *Hazardous and Toxic Effects of Industrial Chemicals,* Noyes Data Corporation, Park Ridge, N.J. 1979.

T. L. van Winkle, J. Edeleanu, E. A. Prosser, and C. A. Walker, "Cotton versus Polyester," *American Scientist,* **66** (May-June 1971).

R. C. Weast, Ed., *CRC Handbook of Chemistry and Physics*, 58th ed., CRC Press Inc., 1977–1978.

A. S. West, "The Impact of TSCA," *Chemical Engineering Progress* (April 1979).

V. Yang and S. C. Trindade, "Brazil's Gasohol Program," *Chemical Engineering Progress* (April 1979).

CHAPTER TWO

Economic Model
of the Intermediate
U.S. Chemical Industry

In this chapter we view the current chemical industry in the United States as a mature, highly developed, and well-integrated industry that responds to change by orderly perturbations from a stable economic base. The current and projected prices of the hundreds of intermediate chemicals are viewed as accurate representations of their value in the economy. Changes in process technology are perceived as attempts to make the industry economically more efficient in response to projected changes in the economy at large.

The economic model of the industry described in this chapter contains 182 processes involved in the transformations of 131 chemical intermediates and feedstocks. Using the economic environments defined by the past, present, and future chemicals prices, supplies, and demands, a comparison is made between this model of the industry and the actual U.S. industry in 1977 as well as the changes required in the structure of the industry to meet projected demands for chemicals in 1985. The behavior of the projected industry is also studied with inclusion of 10 potential new processes currently under development. Improvements needed in the new technologies to make them valuable to the 1985 industry are discussed. We also show how sensitive the projected industry is to the several perturbations such as distortions in energy cost, shortages in propane and natural gas supplies, and use of coal-derived synthesis gas. Finally, the major implications of this type of model of the industry are summarized and the potential applications outlined.

MODEL CONSTRUCTION AND DESCRIPTION

The model of the U.S. chemical industry, in general, is based on the following basic assumptions which are seen in Figure 2.1 and summarized as follows:

1   The chemical industry as the whole can be visualized and represented as a network of chemical processes.

2   The processes involve chemical and/or physical transformation and are therefore defined by their specific material flows. Each process can be thought of as an entry in a technology catalog.

3   The material interactions among processes are linear. The processing network therefore can be represented as a linear input/output matrix.

4   There exists a specific supply/demand environment and process capacity limitations.

In the model, it is assumed the overall industry seeks to utilize its available resources in an optimal fashion. The economic model includes 182 processes involved in the transformation of 131 chemical intermediates and feedstocks. Tables 2.1 and 2.2 list the chemicals and processes included in the model.

Within the specified process capacity limits B_j and feedstock supply constraints S_i, the model identifies processes to operate at levels X_j, such that the product demands D_i are satisfied at a minimum total production cost to the industry.

The production cost to the industry is defined as

$$\text{production cost} = \sum_{i=1}^{N} P_i F_i + \sum_{j=1}^{M} C_j X_j + \sum_{i=1}^{N} (Q_i - D_i)(P_i - H_i)$$

where C_j is the total cost of production for process j and P_i and H_i are chemical prices and heating values for chemical i, respectively. The term C_j is composed of raw material costs M_j, utilities cost U_j, labor-related costs L_j, and an investment-related cost I_j. By-products are valued as credits and are included in the raw material cost. The term $P_i F_i$ represents cost of feedstock materials, and the term $(Q_i - D_i)(P_i - H_i)$ represents a correction term for surplus chemicals. (Any surplus chemicals $Q_i - D_i$ resulting from the model network has heating value instead of chemical value.) See Table 2.3.

In the remainder of this chapter we illustrate the uses of this model as a tool to detect the future trends of this technically complex industry. Tables 2.4, 2.5, and 2.6 contain estimates of the supply of feedstocks, the demands for chemicals, and the industrial capacity for the United States. These data are the basis for the observations of this chapter.

PROCESS SELECTION BY THE MODEL IN 1977

To test the model of the industry, a study run was made based on the actual 1977 supply/demand and existing process capacities. In the study case, the industry was assumed to have the freedom to select the optimal processes to satisfy the product demands within the known capacity limitations. In other

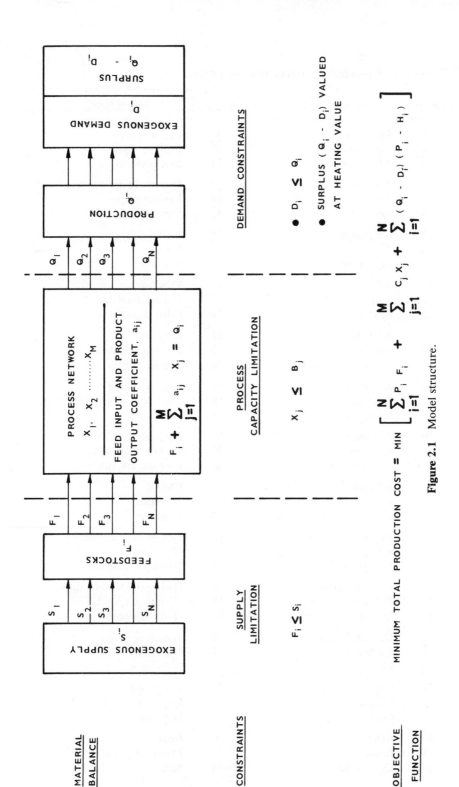

Figure 2.1 Model structure.

Table 2.1 Chemical Products and Feedstocks

Number	Chemical Name	Number	Chemical Name
1	Acetaldehyde	40	Cyclohexanol
2	Acetic acid	41	Cyclohexanone
3	Acetic anhydride	42	Diethylene glycol
4	Acetone	43	Dimethyl terephthalate
5	Acetylene	44	Dinitrotoluene
6	Acrolein	45	Dipropylene glycol
7	Acrylamide	46	Epichlorohydrin
8	Acrylic acid	47	Ethane
9	Acrylonitrile	48	Ethanol
10	Adipic acid	49	Ethyl acetate
11	Adiponitrile	50	Ethyl acrylate
12	Allyl alcohol	51	Ethyl benzene
13	Allyl chloride	52	Ethylene
14	Ammonia	53	Ethylene dichloride
15	Ammonium sulfate	54	Ethylene glycol
16	Aniline	55	Ethylene oxide
17	Benzene	56	2-Ethylhexanol
18	Bisphenol-A	57	Formaldehyde
19	Butadiene	58	Fuel gas
20	n-Butane	59	Fuel oil, high sulfur
21	1,4-Butanediol	60	Fuel oil, low sulfur
22	n-Butanol	61	Gas oil
23	s-Butanol	62	Glycerine
24	Butenes, steam cracked	63	Heptenes
25	t-Butyl alcohol, crude	64	Hexamethylenediamine
26	n-Butylenes	65	Hydrogen
27	Butyraldehyde	66	Hydrogen chloride
28	Calcium carbonate	67	Hydrogen chloride, dilute
29	Calcium oxide	68	Hydrogen cyanide
30	Caprolactam	69	Hydrogen peroxide
31	Carbon dioxide	70	Isobutane
32	Carbon monoxide	71	Isobutanol
33	Chlorine	72	Isobutylene
34	Chlorobenzene	73	Isooctanol
35	Chloroprene	74	Isophthalic acid
36	Coal	75	Isoprene
37	Coke	76	Isopropanol
38	Cumene	77	Maleic anhydride
39	Cyclohexane	78	Melamine

Table 2.1 *(Continued)*

Number	Chemical Name	Number	Chemical Name
79	Methane	106	Pyrolysis gasoline
80	Methanol	107	Sodium carbonate
81	Methyl acetate	108	Sodium chloride
82	Methyl acrylate	109	Sodium hydroxide
83	Methylene diphenylene diisocyanate (MDI)	110	Sodium hydroxide, dilute
		111	Sodium sulfate
84	Methyl ethyl ketone	112	Styrene
85	Methyl isobutyl ketone	113	Sulfur
86	Methyl methacrylate	114	Sulfuric acid
87	Naphtha	115	Synthesis gas (H_2:CO = 1:1)
88	Naphthalene	116	Synthesis gas (H_2:CO = 2:1)
89	Nitric acid, 95%	117	Synthesis gas (H_2:CO = 3:1)
90	Nitric acid, 60%	118	Terephthalic acid, fiber grade
91	Nitrobenzene	119	Terephthalic acid, crude
92	Oleum	120	Toluene
93	Oxygen	121	Toluene diamine
94	Pentane	122	Toluene diisocyanate
95	Pentenes, steam cracked	123	Triethylene glycol
96	Phenol		
97	Phosgene	124	Urea
98	Phthalic anhydride	125	Vinyl acetate
99	Propane (LPG)	126	Vinyl chloride
100	Propylene, chemical grade	127	Vinylidene chloride
101	Propylene, polymer grade	128	*m*-Xylene
102	Propylene, refinery grade	129	*o*-Xylene
103	Propylene dichloride	130	*p*-Xylene
104	Propylene glycol	131	Xylenes (C_8/C_9 aromatics)
105	Propylene oxide		

words, existing process capacities are specified as upper limits. Part of the results are summarized in Tables 2.7 and 2.8 and lead to a few observations.

1　Propane is selected as the most attractive feedstock for producing ethylene. However, to meet the demand for propylene, the model also selects steam cracking of higher-molecular-weight feedstocks such as naphtha and gas oil.

2　Based on the economic environment defined to the model, the process of direct sulfonation of benzene to phenol is selected rather than the more

Table 2.2 Chemical Processes

Process Number	Process Description
1	Acetaldehyde via ethylene (one-step oxidation)
2	Acetaldehyde via ethylene (two-step oxidation)
3	Acetaldehyde via oxidation of ethanol
4	Acetic acid via carbonylation of methanol
5	Acetic acid via air oxidation of acetaldehyde
6	Acetic acid via oxidation of n-butane
7	Acetic acid via oxidation of n-butylenes
8	Acetic anhydride via oxidation of acetaldehyde
9	Acetic anhydride via ketene and acetic acid
10	Acetone via v.p. dehydrogenation of isopropanol
11	Acetone via oxidation of propylene
12	Acetylene from residual oil (submerged-flame process)
13	Acetylene via hydration of calcium carbide
14	Acetylene via pyrolysis of methane
15	Acetylene via pyrolysis of naphtha
16	Acetylene via pyrolysis of ethane
17	Acrolein via oxidation of propylene
18	Acrylamide via hydration of acrylonitrile (fixed-bed catalyst)
19	Acrylamide via hydration of acrylonitrile (suspended catalyst)
20	Acrylamide via sulfuric acid process
21	Acrylic acid via oxidation of propylene
22	Acrylic acid via carbonylation of acetylene
23	Acrylonitrile via ammoxidation of propylene
24	Acrylonitrile via cyanation/oxidation of ethylene
25	Adipic acid from cyclohexane
26	Adipic acid via cyclohexanol
27	Adiponitrile via adipic acid and ammonia
28	Adiponitrile via hydrodimerization of acrylonitrile
29	Allyl alcohol via isomerization of propylene oxide
30	Allyl alcohol via acrolein and s-butyl alcohol
31	Allyl chloride via chlorination of propylene
32	Ammonia from natural gas
33	Ammonia from naphtha
34	Aniline via mononitrobenzene
35	Aniline via ammonolysis of cyclohexanol
36	Aniline via phenol and ammonia

Table 2.2 *(Continued)*

Process Number	Process Description
37	Benzene via hydrodealkylation of toluene
38	Benzene via disproportionation of toluene
39	Bisphenol-A via phenol and acetone
40	Butadiene via dehydrogenation of *n*-butylenes
41	Butadiene from *n*-butylenes (oxidative dehydrogenation)
42	Butadiene via dehydrogenation of *n*-butane
43	1,4-Butanediol via acetylene and formaldehyde
44	1,4-Butanediol from butadiene
45	1,4-Butanediol from propylene, chem. grade
46	1,4-Butanediol via propylene oxide
47	*n*-Butanol via propylene (conventional oxo)
48	*n*-Butanol via propylene (co-phosphine catalyst)
49	*n*-Butanol via propylene (rhodium catalyst)
50	*s*-Butanol via sulfonation of *n*-butylenes
51	Butyraldehyde via oxonation of propylene
52	Caprolactam via hexahydrobenzoic acid
53	Caprolactam via nitric oxide reduction process
54	Caprolactam via phenol process
55	Caprolactam from cyclohexane
56	Caprolactam via cyclohexanone and hydroxylamine
57	Carbon monoxide from natural gas
58	Carbon monoxide from naphtha
59	Chlorine from sodium chloride (electrolysis)
60	Chlorobenzene via chlorination of benzene
61	Chlorobenzene via oxychlorination of benzene
62	Chloroprene via chlorination of butadiene
63	Chloroprene via dimerization of acetylene
64	Cumene from benzene and propylene
65	Cyclohexane via hydrogenation of benzene
66	Cyclohexanol via oxidation of cyclohexane
67	Cyclohexanol from cyclohexane (boric acid process)
68	Cyclohexanol via oxidation of cyclohexane
69	Cyclohexanone via dehydrogenation of cyclohexanol
70	Cyclohexanone via cyclohexane
71	Dimethyl terephthalate from *p*-xylene
72	Dimethyl terephthalate via TPA

Table 2.2 *(Continued)*

Process Number	Process Description
73	Dinitrotoluene via nitration of toluene
74	Epichlorohydrin via allyl chloride
75	Ethanol via hydration of ethylene
76	Ethyl acrylate via acrylic acid
77	Ethyl acrylate via acrylonitrile
78	Ethyl acrylate via acetylene
79	Ethylbenzene via benzene alkylation
80	Ethylene via ethane-propane (50:50) cracking
81	Ethylene via gas oil (high severity) cracking
82	Ethylene via naphtha (high severity) cracking
83	Ethylene via pyrolysis of ethane
84	Ethylene via pyrolysis of propane
85	Ethylene via naphtha (low severity) cracking
86	Ethylene via gas oil (low severity) cracking
87	Ethylene via gas oil (med. severity) cracking
88	Ethylene via hydrogenation of acetylene
89	Ethylene via dehydrogenation of ethanol
90	Ethylene dichloride via chlorination of ethylene
91	Ethylene dichloride via ethylene (oxychlorination)
92	Ethylene glycol via ethylene oxide
93	Ethylene glycol via ethylene oxidation
94	Ethylene oxide via oxidation of ethylene (air)
95	Ethylene oxide via oxidation of ethylene (O_2)
96	Ethylene oxide via chlorohydration of ethylene
97	2-Ethylhexanol via oxo process
98	Formaldehyde via oxidation of methanol
99	Glycerine from allyl chloride
100	Glycerine via epichlorohydrin
101	Glycerine via allyl alcohol and H_2O_2
102	Hexamethylenediamine via acrylonitrile
103	Hexamethylenediamine via adipic acid
104	Hexamethylenediamine via butadiene
105	Hydrogen from methane
106	Hydrogen from naphtha
107	Hydrogen via partial oxidation of naphtha
108	Hydrogen cyanide via ammoxidation of methane
109	Hydrogen peroxide via anthraquinone process
110	Hydrogen peroxide from isopropanol

Table 2.2 *(Continued)*

Process Number	Process Description
111	Isobutane via isomerization of *n*-butane
112	Isobutylene from steam cracked butenes
113	Iso-octanol via heptenes (one-step oxonation)
114	Iso-octanol via heptenes (two-step oxonation)
115	Isophthalic acid from *m*-xylene
116	Isoprene via dimerization of propylene
117	Isoprene via formaldehyde and isobutylene
118	Isoprene from C_5 fractions
119	Isopropanol via hydration of propylene
120	Isopropanol via propylene (cation exchange)
121	Maleic anhydride via oxidation of benzene
122	Maleic anhydride via oxidation of *n*-butane
123	Melamine via BASF Process
124	Melamine via Chemie Linz Process
125	Melamine via Nissan Process
126	Melamine via Stamicarbon Process
127	Methanol from methane
128	Methanol via carbon monoxide (high pressure)
129	Methanol via carbon monoxide (low pressure)
130	Methyl acrylate via esterification of acrylic acid
131	Methylene diphenylene diisocyanate via aniline and phosgene
132	Methyl ethyl ketone via *s*-butanol
133	Methyl ethyl ketone from *n*-butylenes
134	Methyl isobutyl ketone via acetone
135	Methyl methacrylate via acetone cyanohydrin
136	Methyl methacrylate from isobutylene
137	Nitric acid (95%) via ammonia
138	Nitric acid (60%) via ammonia
139	Nitrobenzene via nitration of benzene
140	Phenol via air oxidation of cumene
141	Phenol via dehydrochlorination of chlorobenzene
142	Phenol via alkaline hydrolysis of chlorobenzene
143	Phenol via sulfonation of benzene
144	Phosgene via carbon monoxide and chlorine
145	Phthalic anhydride from *o*-xylene
146	Phthalic anhydride from naphthalene

Table 2.2 *(Continued)*

Process Number	Process Description
147	Propylene, chemical grade from propylene, refinery
148	Propylene, polymer grade from propylene, refinery
149	Propylene, polymer grade via propylene, chemical grade
150	Propylene glycol via hydration of propylene oxide
151	Propylene oxide via chlorohydration of propylene
152	Propylene oxide via oxidation of propylene
153	Styrene via ethylbenzene (dehydrogenation)
154	Styrene via ethylbenzene (hydroperoxide process)
155	Sulfuric acid via double absorption process
156	Synthesis gas (H_2:CO = 1:1) via methane reforming
157	Synthesis gas (H_2:CO = 1:1) from residual oil
158	Synthesis gas (H_2:CO = 2:1) via coal gasification
159	Synthesis gas (H_2:CO = 2:1) from naphtha
160	Synthesis gas (H_2:CO = 2:1) from residual oil
161	Synthesis gas (H_2:CO = 3:1) via coal gasification
162	Synthesis gas (H_2:CO = 3:1) from residual oil
163	Synthesis gas (H_2:CO = 3:1) via methane reforming
164	Terephthalic acid, fiber grade via *p*-xylene
165	Terephthalic acid, fiber grade via crude TPA
166	Terephthalic acid, crude grade via *p*-xylene
167	Terephthalic acid, crude grade via acetaldehyde
168	Toluene diamine via dinitrotoluene
169	Toluene diisocyanate via toluene diamine
170	Urea via ammonia and carbon dioxide
171	Urea via total recycle process
172	Vinyl acetate via ethylene and acetic acid
173	Vinyl acetate via acetylene and acetic acid
174	Vinyl acetate via ethane and acetic acid
175	Vinyl chloride from ethylene
176	Vinyl chloride via ethylene dichloride
177	Vinyl chloride via acetylene
178	Vinylidene chloride via dehydrochlorination of 1,1,2-trichloroethane
179	Vinylidene chloride via vinyl chloride
180	Vinylidene chloride via ethane chlorination
181	*p*-Xylene via isomerization of *m*-xylene (Aromax-Isolene)
182	*p*-Xylene via isomerization of *m*-xylene (Parex-Isomar)

Table 2.3 Estimated Heating Values and Chemical Prices (¢/lb Current $ Basis)

Chemicals	MBTU/lb	Heating Values		Chemical Prices	
		1975	1977	1975	1977
Acetaldehyde	0.01139	1.92	2.28	14.2	16.5
Acetic acid	0.00627	1.05	1.25	13.8	16
Acetic anhydride	0.00761	1.28	1.52	21.6	25
Acetone	0.01326	2.23	2.65	12.4	14.5
Acetylene	0.02147	3.61	4.29	23	25
Acrolein	0.01251	2.10	2.50	26.7	31
Acrylamide	—	0.86	1.05	31.9	37
Acrylic acid	0.00817	1.37	1.63	24.2	28
Acrylonitrile	0.00861	1.05	1.29	22.5	26
Adipic acid	0.00823	1.4	1.65	24.2	28
Adiponitrile	0.00850	1.05	1.28	34.6	40
Allyl alcohol	0.01371	1.46	2.74	30.6	35
Allyl chloride	0.00979	0	0	14.6	17
Ammonia	0.00718	0.88	1.08	6.5	7.5
Ammonium sulfate	0	0	0	2.6	3
Aniline	0.01569	1.92	2.35	23.3	27
Benzene	0.01800	3.03	3.60	9.3	10.7
Bisphenol-A	0.01378	2.71	2.75	34.5	40
Butadiene	0.02022	2.48	3.03	18.7	19.3
n-Butane	0.02129	2.61	3.19	3.8	4.6
Isobutane	0.02124	2.61	3.19	4.2	5.0
1,4-Butanediol	—	2.57	3.05	43.3	50.0
Iso-Butanol	0.01550	2.53	3.10	18.5	21.5
n-Butanol	0.01553	2.53	3.10	15	17.5
sec-Butanol	0.01540	2.52	3.08	13	15
tert-Butanol	0.01528	2.50	3.06	6.5	7.6
Butenes, mixed	0.02076	2.54	3.11	5.3	6.3
n-Butenes	0.02080	2.55	3.12	5.9	7.0
Iso-Butenes	0.02072	2.54	3.11	11.3	13
Butyraldehyde	0.01322	2.22	2.64	22.46	26
Calcium carbonate	0	0	0	0.87	1
Calcium oxide	0	0	0	1.5	1.7
Caprolactam	0.0125	1.54	1.88	46.6	54
Carbon dioxide	0	0	0	0.25	0.3
Carbon monoxide	0.00434	0.53	0.65	2.1	2.5
Chlorine	0	0	0	5.4	6.0

Table 2.3 *(Continued)*

Chemicals	MBTU/lb	Heating Values 1975	Heating Values 1977	Chemical Prices 1975	Chemical Prices 1977
Chlorobenzene	0.01260	0	0	22.7	26.0
Chloroprene	0.005	0	0	26.2	30
Coal (Wyoming)	0.0085	0.57	0.68	0.57	0.68
Coal (Illinois)	0.0115	0.78	0.93	0.78	0.93
Coke	0.01409	0.94	1.12	2.6	3
Cumene	0.01868	3.15	3.74	11.3	13
Cyclohexane	0.02053	3.45	4.10	12.1	14
Cyclohexanol	0.01600	2.69	3.20	25	29
Cyclohexanone	0.01600	2.69	3.20	26.8	31
Diethylene glycol	0.00927	1.56	1.85	17.8	20.5
DMT	0.01038	1.74	2.07	26.8	31
Dinitrotoluene	0.00843	1.03	1.26	12.1	14
Dipropylene glycol	0.0113	1.90	2.26	23.4	27
Epichlorohydrin	0.006	0	0	26	30
Ethane	0.02230	3.75	4.46	4.33	5.15
Ethanol (95%)	0.01276	2.08	2.55	14.5	17
Ethyl acetate	0.01097	1.79	2.19	19.6	22.5
Ethyl acrylate	0.01287	2.10	2.57	28.5	33
Ethylbenzene	0.01850	3.11	3.70	11.3	13
Ethylene	0.02164	3.54	4.33	10.6	12.47
Ethylene dichloride	0.00493	0	0	8.6	10
Ethylene glycol	0.00818	1.34	1.64	19.5	22.5
Ethylene oxide	0.01234	2.02	2.47	20.7	24
2-Ethylhexanol	0.01744	2.94	3.49	20	23
Formaldehyde (100%)	0.00818	1.38	1.64	12.2	14
Fuel gas	0.0500	6.1	7.5	6.1	7.5
Fuel oil, HS	0.01857	2.32	2.84	2.32	2.84
Fuel oil, LS	0.01886	3.17	3.77	3.17	3.77
Gas oil/distillate	0.02076	3.49	4.15	4.26	5.0
Glycerine	0.00776	1.27	1.55	40.7	47
Heptene	0.02059	3.37	4.12	10.4	12
Hexamethylenediamine	0.01645	2.02	2.47	41.5	48
Hydrogen	0.06110	7.50	9.17	21	25
HCl	0	0	0	5.37	6.2
HCl, dilute (30%)	0	0	0	1.5	1.7
HCN	0.00974	1.19	1.46	21.7	25

Table 2.3 *(Continued)*

Chemicals	MBTU/lb	Heating Values		Chemical Prices	
		1975	1977	1975	1977
Hydrogen peroxide	0	0	0	35.3	41
Isooctanol	0.01744	2.94	3.49	19	22
Isophthalic acid	0.00835	1.36	1.67	24.2	28
Isoprene	0.02060	2.53	3.09	21.8	25
Isopropanol	0.01422	2.32	2.84	11	13
Maleic anhydride	0.0061	1.03	1.22	26	30.0
Melamine	—	0.49	0.60	30.57	35
Methane	0.0250	3.06	3.75	3.6	4.5
Methanol	0.00976	1.59	1.95	5.9	7.0
Methyl acetate	0.00926	1.51	1.85	5.15	6.0
Methyl acrylate	0.012	1.47	1.80	29.26	34
Methylene diphenylene diisocyanate (MDI)	0.014	1.72	2.1	42.23	48.8
Methyl ethyl ketone	0.01458	2.46	2.92	16	18.5
Methyl isobutyl ketone	0.01609	2.71	3.22	23.3	27.0
Methyl methacrylate	0.010	1.68	2.00	35.3	41
Naphtha	0.0205	3.35	4.10	5.0	6.0
Naphthalene	0.01730	2.83	3.46	11.3	13
Nitric acid (95%)	0	0	0	5.2	6.0
Nitric acid (60%)	0	0	0	3.9	4.5
Nitrobenzene	0.01081	1.32	1.62	12.6	14.6
Oleum	0	0	0	2.3	2.6
Oxygen	0	0	0	1.35	1.6
Pentane	0.02107	2.58	3.16	6.0	7.0
Pentenes	0.02060	2.52	3.09	5.4	6.3
Phenol	0.01395	2.28	2.79	20	23
Phosgene	0.001	0	0	9.2	10.5
Phthalic anhydride	0.00954	1.61	1.91	20.5	24
Propane	0.02165	2.66	3.25	4.2	5
Propylene, chemical grade	0.02104	2.58	3.16	7	8.5
Propylene, polymer grade	0.02104	2.58	3.16	7.78	9.5
Propylene, refinery grade	0.02104	2.58	3.16	6.3	7.5
Propylene dichloride	0.005	0	0	13	15
Propylene glycol	0.01021	1.67	2.04	20	23
Propylene oxide	0.01437	2.35	2.87	19.5	22.5
Pyrolysis gasoline	0.01950	3.28	3.90	4.9	5.8

Table 2.3 *(Continued)*

Chemicals	MBTU/lb	Heating Values 1975	Heating Values 1977	Chemical Prices 1975	Chemical Prices 1977
Sodium carbonate	0	0	0	4.2	4.8
Sodium chloride	0	0	0	1.4	1.6
Sodium hydroxide	0	0	0	6.7	7.5
Sodium hydroxide, dilute	0	0	0	1.8	2.0
Sodium sulfate	0	0	0	2.2	2.5
Styrene	0.0181	3.05	3.62	16.7	19.5
Sulfur	0.0053	0	0	2.4	2.7
Sulfuric acid	0	0	0	1.95	2.25
Synthesis gas (H_2:CO = 1:1)	0.00814	1.00	1.22	3.0	3.5
Synthesis gas (H_2:CO = 2:1)	0.01144	1.41	1.72	4.6	5.3
Synthesis gas (H_2:CO = 3:1)	0.01436	1.76	2.15	5.2	6.0
Terephthalic acid, fiber grade	0.00835	1.41	1.67	25.9	30
Terephthalic acid, crude	0.00835	1.41	1.67	16.3	19
Toluene	0.01825	3.07	3.65	6.0	6.9
Toluene diamine	0.01405	1.72	2.11	31.1	36
Toluene diisocyanate	0.012	1.47	1.80	43.2	50
Triethylene glycol	0.01019	1.72	2.04	30.3	35
Urea	0.00452	0.57	0.68	6.7	7.8
Vinyl acetate	0.01031	1.73	2.06	19.4	23
Vinyl chloride	0.00818	0	0	13	15
Vinylidene chloride	—	0	0	23.45	27
m-Xylene	0.01845	3.10	3.69	18.5	22
o-Xylene	0.01850	3.11	3.70	9.3	11.1
p-Xylene	0.01847	3.11	3.69	12.8	15.2
Xylenes	0.01848	3.29	3.70	5.9	6.9

Table 2.4 Actual/Projected U.S. Exogenous Supply[1,2] of Chemicals (Millions of Pounds)

Chemicals	1975	1977	1980[3]	1985[3]
Acetaldehyde	25	25	25	25
Acetic acid	175	175	175	175
Acetone	3	15	15	15
Acrylic acid	10	0	0	0
Adipic acid	85	25	25	25
Allyl alcohol	40	40	40	40

Table 2.4 *(Continued)*

Chemicals	1975	1977	1980[3]	1985[3]
Ammonia	910	1,130	3,850	3,860
Aniline	50	50	60	70
Benzene	8,950	10,360	11,700	12,500
Butadiene	1,530	2,200	2,750	3,150
n-Butane	41,500	43,660	46,900	52,250
Butenes	20,800	21,700	23,800	28,400
t-Butanol, crude	12	12	12	12
Cyclohexane	330	330	330	330
Cyclohexanol	225	235	250	280
Cyclohexanone	400	420	450	480
Ethane	17,050	17,600	18,470	20,000
Ethanol	75	75	75	75
Ethyl acetate	265	265	265	265
Ethylbenzene	420	460	460	460
Ethylene	3,400	3,600	4,900	5,400
Heptenes	125	125	125	125
Hexamethylenediamine	75	85	100	130
Hydrogen cyanide	290	320	370	470
Isobutane	43,100	45,600	49,400	55,700
Isobutylenes	360	480	630	800
Isopropanol	50	50	0	0
Methyl ethyl ketone	175	175	175	175
Naphthalene	760	830	950	1,130
Pentenes, steam cracked	600	870	1,500	3,000
Phenol	55	55	55	55
Propylene, refinery grade	4,500	4,150	4,400	4,700
Sodium carbonate	17,350	17,650	18,200	19,100
Toluene	10,400	11,600	12,800	15,000
m-Xylene	150	150	200	350
o-Xylene	1,350	1,350	1,370	1,400
p-Xylene	1,350	1,360	2,150	2,450
Xylenes	9,400	10,150	12,000	15,000

[1] Imports are excluded.

[2] Exogenous supply of adiponitrile, allyl chloride, ammonium sulfate, calcium carbonate, calcium oxide, carbon dioxide, coal, coke, dipropylene glycol, fuel gas, fuel oil (high and low sulfur), gas oil, hydrogen chloride (pure and dilute), isobutanol, methane, methyl acetate, naphtha, oleum, oxygen, pentane, propane, propylene dichloride, pyrolysis gasoline, sodium chloride, sodium hydroxide (pure and dilute), sodium sulfate, and sulfur are not constrained.

[3] Exogenous supply of chemicals for 1980 and 1985 are estimated based on the supply/demand patterns for the past several years.

Table 2.5 Actual/Projected U.S. Exogenous Demand[1] for Petrochemicals (Millions of Pounds)

Chemicals	1975	1977	1980[2]	1985[2]
Acetaldehyde	365	420	500	550
Acetic acid	800	870	900	1,000
Acetic anhydride	1,700	1,800	1,870	2,050
Acetone	900	990	1,000	1,200
Acetylene	180	180	230	250
Acrolein	35	35	35	35
Acrylamide	50	75	85	125
Acrylic acid	165	250	285	400
Acrylonitrile	825	1,130	1,280	1,600
Adipic acid	1,130	1,430	1,660	2,150
Ammonia	19,860	22,850	22,450	23,300
Ammonium sulfate	4,490	4,670	4,800	5,560
Aniline	235	270	290	360
Benzene	415	445	490	570
Bisphenol-A	300	455	600	1,000
Butadiene	2,860	3,600	3,800	4,500
n-Butane[3]	1,500	1,500	1,500	1,500
1,4-Butanediol	175	215	280	390
n-Butanol	450	500	600	760
s-Butanol	10	10	20	20
Butenes	450	575	750	1,100
n-Butylenes[3]	65	80	95	120
Caprolactam	630	830	910	1,160
Chlorine	8,450	9,200	10,600	13,500
Chlorobenzene	255	350	350	360
Chloroprene	360	400	460	600
Cumene	85	105	125	150
Cyclohexane	30	35	35	35
Cyclohexanol	70	75	80	85
Cyclohexanone	400	400	410	450
Diethylene glycol	385	400	470	600
Dimethyl terephthalate	2,670	2,930	3,300	3,950
Epichlorohydrin	210	225	240	275
Ethanol	910	910	1,000	1,200
Ethyl acetate	145	145	145	145
Ethyl acrylate	230	450	600	960
Ethylbenzene	45	60	60	60
Ethylene	9,100	12,050	13,860	19,600

Table 2.5 *(Continued)*

Chemicals	1975	1977	1980[2]	1985[2]
Ethylene dichloride	1,825	1,950	2,150	2,900
Ethylene glycol	3,480	3,575	4,150	5,200
Ethylene oxide	1,600	1,800	2,300	3,000
2-Ethylhexanol	375	500	585	800
Formaldehyde (100%)	1,860	2,035	2,300	2,600
Glycerine	300	300	320	340
Heptenes	10	40	30	20
Hexamethylenediamine	750	870	1,050	1,350
Hydrogen cyanide	90	100	120	150
Hydrogen peroxide	180	220	290	470
Isobutane[3]	0	0	0	0
Isobutanol	145	145	150	150
Isobutylenes[3]	420	600	660	770
Isooctanol	10	10	15	15
Isophthalic acid	110	140	200	350
Isoprene	260	290	350	400
Isopropanol	815	885	950	1,100
Maleic anhydride	215	290	350	500
Melamine	115	130	150	200
Methanol	2,120	2,825	3,660	4,400
Methyl acrylate	40	45	55	70
Methylene diphenylene diisocyanate	195	240	370	690
Methyl ethyl ketone	375	500	590	770
Methyl isobutyl ketone	140	170	140	120
Methyl methacrylate	470	650	900	1,300
Naphthalene	165	180	200	250
Nitrobenzene	15	20	25	25
Oxygen[4]	520	535	550	600
Phenol	1,100	1,400	1,650	2,100
Phosgene	285	360	510	720
Phthalic anhydride	755	980	1,100	1,400
Propylene, polymer grade	3,100	4,470	5,300	8,250
Propylene glycol	395	465	530	720
Propylene oxide	1,020	1,360	1,650	2,170
Sodium carbonate	12,770	12,940	13,350	14,000
Sodium chloride	14,500	16,000	18,500	23,600
Sodium sulfate	3,700	3,900	4,300	5,000
Styrene	4,385	5,675	6,780	9,100
Sulfur	2,770	2,765	2,750	2,550

Table 2.5 *(Continued)*

Chemicals	1975	1977	1980[2]	1985[2]
Terephthalic acid, fiber grade	975	1,375	1,975	2,970
Toluene	1,890	3,350	4,400	6,700
Toluene diisocyanate	575	620	700	850
Triethylene glycol	105	110	130	170
Urea	8,070	9,075	11,950	19,000
Vinyl acetate	1,080	1,275	1,650	2,600
Vinyl chloride	3,800	5,460	6,670	9,050
Vinylidene chloride	180	210	260	360
o-Xylene	15	15	20	25
Xylenes	1,730	2,330	3,100	5,250

[1]Exports are excluded.
[2]Exogenous demand of chemicals for 1980 and 1985 are estimated based on the supply/demand patterns for the past several years.
[3]Demand for gasoline blending is excluded.
[4]Demand for iron manufacturing is excluded.

Table 2.6 Actual/Projected U.S. Process Capacities (Millions of Pounds)

Chemicals	Processes[1]	1975	1977	1980[2]	1985[2]
Acetaldehyde	1	500	550	—[3]	—[3]
	3	185	85	85	85
Acetic acid	4	490	615	1,900	1,900
	5	930	1,230	1,230	1,230
	6	1,150	1,250	1,250	1,250
Acetone	10	970	970	1,300	1,300
	11	40	40	—[3]	—[3]
Acetylene	12	0	0	—[3]	—[3]
	13	575	575	575	575
	14	435	435	450	450
	16	65	65	65	65
Acrolein	17	90	90	90	90
Acrylamide	18	30	70	—[3]	—[3]
	20	55	55	—[3]	—[3]
Acrylic acid	21	380	740	—[3]	—[3]
	22	350	350	—[3]	—[3]

Table 2.6 *(Continued)*

Chemicals	Pro-cesses[1]	1975	1977	1980[2]	1985[2]
Acrylonitrile	23	1,615	1,860	2,300	2,700
Adipic acid	25	260	490	600	800
	26	1,280	1,335	1,400	1,600
Adiponitrile	27	180	205	250	300
	28	150	175	200	250
Allyl alcohol	30	50	50	50	50
Allyl chloride	31	440	465	500	600
Ammonia	32	34,600	42,800	—[3]	—[3]
Aniline	34	585	695	1,100	—[3]
Benzene	37	4,100	4,500	5,400	5,400
	38	90	180	200	200
Bisphenol-A	39	535	535	850	—[3]
Butadiene	40	530	275	275	275
	41	1,070	1,070	1,400	1,400
	42	1,200	860	500	500
1,4-Butanediol	43	240	425	500	650
s-Butanol	50	490	585	800	800
Caprolactam	53	470	650	—[3]	—[3]
	54	400	420	—[3]	—[3]
Chlorine	59	25,600	29,300	36,000	50,500
Chlorobenzene	60	325	360	360	360
	61	325	360	360	360
Chloroprene	62	400	440	500	650
Cumene	64	3,655	4,735	5,200	5,300
Cyclohexane	65	2,240	2,650	2,800	—[3]
Cyclohexanol	66	960	1,000	1,100	1,200
Cyclohexanone	69	1,105	1,150	1,200	1,350
	70	70	75	80	100
Dimethyl terephthalate	71	3,565	3,905	4,400	5,300
	72	595	655	750	900
Dinitrotoluene	73	1,040	1,040	1,040	—[3]
Epichlorohydrin	74	450	475	500	600
Ethanol	75	2,115	2,115	2,115	2,115

Table 2.6 *(Continued)*

Chemicals	Processes[1]	1975	1977	1980[2]	1985[2]
Ethylbenzene	79	8,125	9,870	—[3]	—[3]
Ethylene	80	6,440	8,375	11,000	14,000
	81	1,720	4,100	6,000	7,400
	82	2,800	3,730	5,800	7,500
	83	855	1,110	1,500	1,900
	84	18,130	19,435	23,000	23,000
	85	2,800	3,730	5,800	7,500
	86	1,720	4,100	6,000	7,400
	87	1,720	4,100	6,000	7,400
Ethylene dichloride	90	14,100	15,070	—[3]	—[3]
	91	14,100	15,070	—[3]	—[3]
Ethylene glycol	92	4,805	5,095	6,000	6,200
Ethylene oxide	94	1,800	1,980	2,300	4,100
	95	3,030	3,685	5,400	4,100
	96	100	100	100	100
2-Ethylhexanol	97	450	555	—[3]	—[3]
Formaldehyde	98	3,270	3,270	3,300	—[3]
Glycerine	99	225	225	250	250
	100	225	225	250	250
	101	90	90	100	100
Hexamethylenediamine	102	150	175	200	250
	103	180	205	250	300
	104	545	630	750	950
Hydrogen cyanide	108	280	305	350	450
Hydrogen peroxide	109	155	190	300	500
	110	85	100	100	130
Isobutylene	112	780	1,200	1,900	2,000
Isoprene	116	60	35	—[3]	—[3]
	118	245	275	350	350
Isopropanol	120	2,210	2,320	2,800	2,800
Maleic anhydride	121	345	400	—[3]	—[3]
	122	0	90	—[3]	—[3]
Melamine	126	170	170	170	250
Methanol	127	8,630	11,380	15,000	21,000
Methyl ethyl ketone	132	415	460	750	750

Table 2.6 (Continued)

Chemicals	Processes[1]	1975	1977	1980[2]	1985[2]
Methyl isobutyl ketone	134	245	245	245	245
Methyl methacrylate	135	795	795	—[3]	—[3]
	136	0	0	—[3]	—[3]
Phenol	140	2,610	2,810	3,300	3,900
	142	50	—[3]	—[3]	—[3]
	143	135	175	175	175
Phosgene	144	1,500	1,800	2,300	3,000
Phthalic anhydride	145	760	835	950	1,300
	146	245	385	385	385
Propylene, chemical grade	147	980	1,045	1,250	1,250
Propylene, polymer grade	148	2,070	2,070	2,070	2,070
	149	2,955	3,755	—[3]	—[3]
Propylene glycol	150	665	665	—[3]	—[3]
Propylene oxide	151	1,605	1,685	—[3]	—[3]
	152	920	920	—[3]	—[3]
Styrene	153	6,780	8,120	10,200	14,000
	154	0	1,000	1,000	1,000
Terephthalic acid, fiber grade	164	1,490	2,100	3,000	4,500
Terephthalic acid, crude	166	525	575	650	800
Toluene diamine	168	626	625	625	—[3]
Toluene diisocyanate	169	825	825	—[3]	—[3]
Urea	170	1,540	1,810	2,300	3,500
	171	8,740	10,270	13,000	19,600
Vinyl acetate	172	1,800	1,800	2,000	2,500
	173	210	210	250	300
Vinyl chloride	175	6,455	6,805	9,200	9,000
	177	425	300	300	300
p-Xylene	181	1,240	1,990	2,000	2,500
	182	340	1,135	1,800	1,800

[1]The process numbers refer to Table 2.2.
[2]Production capacities for 1980 and 1985 are estimated based on the supply and demand patterns of intermediate chemicals.
[3]Not constrained.

Table 2.7 Model vs. Actual Chemical Production in 1977 (Millions of Pounds)

Chemical Production	Model[1]	Actual[2]
Acetone	1,765	2,100
Acrylic acid	650	550
Acrylonitrile	1,400	1,400
Adipic acid	1,550	1,600
Butadiene	4,100	3,800
Caprolactam	830	830
Ethylene	24,000	24,400
Maleic anhydride	290	300
Methyl ethyl ketone	500	550
Methyl isobutyl ketone	170	150
Phenol	1,810	2,300
Phthalic anhydride	980	1,000
Propylene glycol	460	460
Propylene oxide	1,800	1,850
Styrene	5,675	5,800
Vinyl chloride	5,500	5,500

[1]Total production within the network + exogenous supply.
[2]Total production + (imports − exports).

modern cumene-based technology. Environmental costs associated with the sulfonation route may have been understated in leading to this choice.

3 The model industry therefore prefers to produce acetone from either oxidation of propylene or as the side-product of the isopropanol-based hydrogen peroxide.

4 Under the conditions chosen in the model, production of acrylic acid via carbonylation of acetylene would be an attractive route. However the choice depends upon the relative costs and availability of acetylene as an alternative for propylene oxidation route.

5 If hydrogen cyanide was available in larger quantities, the butadiene route to hexamethylenediamine (HMDA) would have been chosen by the model industry. However, to meet the demand for HMDA, acrylonitrile and adipic acid are also used as alternate feedstocks for HMDA.

6 Styrene from ethylbenzene by the hydroperoxide process is very attractive because of the propylene oxide by-product. But due to the limitations in installed capacity the model selects chlorohydration and/or oxidation of propylene to propylene oxide and dehydrogenation of ethylbenzene to styrene to satisfy the demands for these chemicals.

7   Removing capacity constraints represents an ideal case in which industry has the total freedom to select the optimal technologies in order to minimize the total production cost. By doing so, acetic acid, butadiene, and acetylene turn out to be the best feedstocks for production of acetic anhydride, hexamethylenediamine, and acrylic acid, respectively. As a result of these changes, ethylene consumption would drop by about 7% and acetylene also would become an attractive feedstock for production of vinyl chloride.

Table 2.8   Model vs. Actual Process Selection in 1977 (Millions of Pounds)

Process Selection	Model	Actual
Acteone		
1. Via isopropanol-based processes	750	750
2. By-product of phenol from cumene	965	1300
3. Via propylene and others	50	50
Acrylic acid		
1. Via propylene	300	500
2. Via acetylene	350	50
Adipic acid		
1. Via cyclohexane-based processes	1525	1575
2. Others	25	25
Caprolactam		
1. Via toluene-based process	830	—
2. Via cyclohexane-based processes	—	430
3. Via phenol-based process	—	400
Maleic anhydride		
1. Via benzene	200	250
2. Via n-butane	90	30
3. Others	—	20
Phenol		
1. Via cumene	1580	2200
2. Via benzene	175	150
3. Others	55	—
4. Net exports	—	50
Phthalic anhydride		
1. Via o-xylene	595	650
2. Via naphthalene	385	300
3. Net imports	—	50
Vinyl chloride		
1. Via ethylene-based processes	5500	5650
2. Via acetylene	—	250
3. Net exports	—	400

PLANNING THE 1985 U.S. CHEMICAL INDUSTRY

Inefficiencies exist in the current industry in the form of insufficient or excess process capacity and plants utilizing inferior technology. The detection of these inefficiencies might reduce the decision-making errors in long-range planning.

To study the future trends and the directional changes of the industry, to identify process bottlenecks, and to determine the process capacity expansion needs for the 1985 U.S. chemical industry, the industry is allowed to expand beyond its 1977 built capacities which are specified as lower bounds in the model. The existing capacity utilization factor (U.F.) is fixed at 100, 50, and 0% of 1977 capacities. The 100% U.F. reflects a situation where expansions are kept to a minimum, because all the existing capacities are assumed to be utilized fully. The 50% U.F. assumes that at least half of the built capacities will be used, but the remaining 50% can be replaced by alternate processes if they prove to be more economical to the industry. The 0% corresponds to an idealized situation where all the existing plants can be replaced by alternate processes, if necessary, to achieve minimum total production cost to the whole industry.

Under the economic environment defined earlier, the 1985 capacity expansion needs indicated by the model are summarized in Table 2.9. Table 2.10 lists the utilization factors (fraction of the 1977 built capacities to be utilized in 1985) of the processes recommended by the model.

Analyzing the behavior of the industry under different utilization factors leads to the following observations:

1 At 100% U.F., the existing capacities for older processes limit the amount of newer technologies. Operating these older processes at 100% U.F., in fact, triggers capacity expansions of other older processes, thus further hampering the overall industry's efficiency. This results in having surplus capacities for acetone, 1,4-butanediol, cyclohexanone, cumene, dimethyl terephthalate, ethylene, formaldehyde, methanol, methyl isobutyl ketone, propylene oxide, and styrene. This surplus vanishes when utilization factor is reduced to 50%.

2 Operating industry at 100% and 50% utilization factors results in extensive production of ethylene, as a result the model favors acetaldehyde over acetic acid to produce acetic anhydride. By eliminating process capacity constraints (0% U.F.), production of ethylene drops more than 6% (using 50% U.F. as the basis for comparison) and acetic acid becomes the only feed for acetic anhydride production.

3 Under the conditions chosen in the model, oxidation of propylene to acrylic acid is one of the processes which is kept by the model industry when it is utilized at 100% U.F. But when the utilization factor decreases, acetylene becomes an attractive feedstock and finally at 0% U.F., carbonylation of acetylene to acrylic acid replaces the propylene route.

Table 2.9 The U.S. Process Capacity Expansion Needs in 1985 When the Industry Expands beyond 0%, 50%, and 100% of the 1977 Built Capacities (Millions of Pounds)

Chemicals	Processes[1]	Capacity Expansion Needs		
		U.F. = 0%	U.F. = 50%	U.F. = 100%
Acetaldehyde	1	0	2,190	1,455
	3	0	0	300
Acetic acid	4	0	430	0
	6	4,320	0	0
Acetic anhydride	8 + 9	480	480	480
Acetone	11	365	0	0
Acetylene	12	65	10	0
	16	965	395	0
Acrylamide	18	55	30	0
Acrylic acid	22	850	480	110
Adipic acid	25	1,635	1,250	870
Allyl alcohol	30	0	0	5
Aniline	34	140	140	0
Bisphenol-A	39	465	465	465
Butadiene	41	1,065	490	0
s-Butanol	50	0	0	15
Caprolactam	53 + 55 + 56	0	0	90
Chlorobenzene	60 + 61	30	30	30
Cyclohexane	65	0	0	750
Cyclohexanol	66 + 67 + 68	0	0	1,050
Dimethyl terephthalate	71	45	0	0
Dinitrotoluene	73	35	35	35
Ethylbenzene	79	100	135	200
Ethylene	81 + 86 + 87	0	0	12,200
	82 + 85	17,200	16,150	0
Ethylene glycol	92	640	640	640
Ethylene oxide	95	4,300	3,250	2,220
2-Ethylhexanol	97	245	245	245
Glycerine	99 + 100	115	70	25
Hexamethylenediamine	104	570	390	190
Hydrogen cyanide	108	100	670	230
Hydrogen peroxide	109	0	100	220
	110	370	100	0
Isoprene	118	125	110	0
Isopropanol	119 + 120	0	0	110
Maleic anhydride	122	410	210	10
Methyl methacrylate	136	1,300	900	505
Phenol	140	0	0	555
	143	2,400	1,380	0
Phosgene	144	570	570	570
Phthalic anhydride	146	480	480	170

Table 2.9 *(Continued)*

Chemicals	Processes[1]	Capacity Expansion Needs		
		U.F. = 0%	U.F. = 50%	U.F. = 100%
Propylene, polymer grade	148	1,235	750	260
	149	1,200	1,710	2,240
Propylene glycol	150	55	55	55
Styrene	154	6,720	3,075	0
Terephthalic acid, fiber grade	164	870	870	870
Toluene diamine	168	20	20	20
Toluene diisocyanate	169	25	25	25
Urea	170	17,800	12,650	7,500
Vinyl acetate	172 + 173	0	0	590
Vinyl chloride	175 + 176	2,250	2,100	1,950
p-Xylene	182	900	0	0

[1] The process numbers refer to Table 2.2.

4 When the utilization factor approaches zero, oxidation of propylene to acetone and autoxidation of isopropanol to hydrogen peroxide (resulting in acetone as side-product) become the best processes to produce acetone; as a result cumene consumption in the model drops by more than 86%.

5 When the utilization factor declines, acetylene becomes a much better feedstock than butadiene for the production of chloroprene, and butadiene turns out to be the only feed for hexamethylenediamine.

The analysis of different capacity utilization factors is useful in understanding some of the inefficiencies of the current industry and in highlighting possible ways of reducing these inefficiencies.

ACCEPTANCE OF NEW TECHNOLOGIES BY THE U.S. CHEMICAL INDUSTRY

In the course of the last few decades, new technologies have been introduced into the chemical industry to improve competitive positions and to create new markets. The acceptance of new technologies by the industry can be studied by including them in the technology catalog. The model can then select the new technology that is most efficient in meeting the economic objectives of the industry.

Ten potential new processes introduced to the model are shown in Table 2.11. To test the attractiveness of these new processes in the 1985 U.S. chemical in-

Table 2.10   Recommended Utilization Factors of the Processes in 1985 When the Industry Expands beyond 0% and 50% of the 1977 Built Capacities

Chemicals	Processes[1]	Utilization Factors	
		U.F. = 0%	U.F. = 50%
Acetaldehyde	1	93	—[2]
Acetic acid	6	—[2]	67
Acrolein	17	38	61
Acrylonitrile	23	91	50[3]
Ammonia	32	80	80
Benzene	37	15	50[3]
s-Butanol	50	4	52
Caprolactam	53 + 55 + 56	0	86
Chlorine	59	75	80
Chloroprene	62	19	50[3]
Cumene	64	14	50[3]
Cyclohexane	65	52	89
Cyclohexanol	66 + 67 + 68	0	92
Dimethyl terephthalate	71	—[2]	93
Epichlorohydrin	74	58	58
Ethanol	75	75	77
Ethylene	84	55	50[3]
Ethylene dichloride	90 + 91	20	20
Formaldehyde	98	83	88
Isopropanol	119 + 120	94	91
Methanol	127	86	93
Methyl isobutyl ketone	134	48	50[3]
Phenol	140	14	57
Phthalic anhydride	145	63	63
Styrene	153	17	62
p-Xylene	181	0	74

[1]The process numbers refer to Table 2.2.
[2]Expansion needed.
[3]Forced into the industry.

dustry, the existing capacity utilization factors are again fixed at 100, 50, and 0% of 1977 built capacities. Table 2.12 shows the processes accepted by the 1985 industry at the different utilization factors.

The major impact of the acceptance of new technologies on the 1985 industry is shown in Tables 2.13, 2.14, and 2.15, and summarized as follows:

1   Acceptance of the new aniline process based on benzene reduces nitrobenzene production by 58% when the industry operates at 50% U.F. Eliminating capacity bounds (0% U.F.), makes this new process even more attrac-

tive. As a result, nitrobenzene consumption drops more than 97%, just enough to satisfy the exogenous demand of this chemical. At 100% U.F., toluene-based aniline becomes more attractive, resulting in about 4% reduction in phenol production in the model.

2 The new maleic anhydride process based on n-butenes is very desirable in all cases. When the industry is forced to operate at 100% of the 1977 built capacities, the new process would be demand-limited and it would share

Table 2.11 List of the Potential New Processes Introduced to the 1985 U.S. Petrochemical Industry

Process Number	Process Description
N1	Aniline via direct amination of benzene
N2	Aniline from toluene via benzoic acid
N3	Ethylene glycol from synthesis gas and formaldehyde
N4	Ethylene glycol via direct hydration process
N5	Maleic anhydride from n-butenes
N6	Phenol from toluene
N7	Styrene via dimerization of butadiene
N8	Styrene from toluene and carbon monoxide
N9	Styrene via toluene dimerization and disproportionation of stilbene
N10	Vinyl acetate via carbonylation of acetic acid

Table 2.12 New Technologies Accepted in 1985 When the Industry Expands beyond 0, 50, and 100% of the 1977 Built Capacities

Chemical	Process[1]	Capacity Utilization Factors		
		0%	50%	100%
Aniline	N1	X	X	
	N2		X	X
Ethylene glycol	N3			
	N4			
Maleic anhydride	N5	X	X	X
Phenol	N6	X	X	
Styrene	N7			
	N8	X	X	
	N9			
Vinyl acetate	N10	X	X	X

[1]The process numbers refer to Table 2.11.

only 2% of the total production of maleic anhydride. But as the utilization factor is reduced, this process, which is more attractive, would be selected over the old *n*-butane route.

3   Based on the economic environment defined to the model, the direct sulfonation of benzene was the optimum process to phenol. However, the new phenol process based on toluene is more attractive. Therefore at a utilization factor of 0%, toluene becomes the most attractive feedstock for production of phenol, resulting in 77% reduction in consumption of cumene and further expansion of the propylene process to acetone which is required to meet the demand.

Table 2.13 Impact of New Processes on the U.S. Process Capacity Expansion Needs in 1985 When the Industry Expands beyond 0, 50, and 100% of the 1977 Built Capacities (Millions of Pounds)

Chemicals	Processes[1]	Capacity Expansion Needs		
		U.F. = 0%	U.F. = 50%	U.F. = 100%
Acetaldehyde	1	0	1,260	0
	3	0	0	220
Acetic acid	4	0	0	0
	6	2,240	0	0
Acetone	11	595	0	0
Acetylene	16	1,020	395	0
Aniline	34	0	0	0
	N1	835	445	0
	N2	0	50	140
Butadiene	41	1,290	620	0
Ethylbenzene	79	0	0	200
Ethylene	82 + 85	17,100	15,900	0
Maleic anhydride	122	0	0	0
	N5	500	255	10
Methanol	127	0	265	1,140
Phenol	140	0	0	410
	143	0	0	0
	N6	2,875	1,425	0
Propylene oxide	152	0	115	0
Styrene	154	5,065	1,275	0
	N8	3,025	2,760	0
Vinyl acetate	172 + 173	0	0	0
	N10	2,600	2,600	2,600
Vinyl chloride	175 + 176	2,250	2,100	1,300
	177	0	0	650
p-Xylene	182	900	0	0

[1]The process numbers refer to Tables 2.2 and 2.11.

Table 2.14 Impact of New Processes on the Capacity Utilization Factors of the Processes in 1985 When the Industry Expands beyond 0 and 50% of the 1977 Built Capacities

Chemicals	Processes[1]	Utilization Factors	
		U.F. = 0%	U.F. = 50%
Acetaldehyde	1	79	—[2]
Acetic anhydride	8 + 9	81	81
Benzene	37	0	50[3]
Chlorine	59	74	80
Chloroprene	62	0	50[3]
Cumene	64	3	50[3]
Ethylbenzene	79	65	68
Ethylene	84	50	50[3]
Maleic anhydride	122	0	50[3]
Methanol	127	100	—[2]
Phenol	140	0	57
	143	0	50[3]
Propylene oxide	152	35	—[2]
Styrene	153	0	50[3]
p-Xylene	181	0	50
	182	—[2]	90

[1]The process numbers refer to Table 2.2.
[2]Expansion needed.
[3]Forced into the industry.

Table 2.15 Impact of Acceptance of the New Processes on the Total Consumption of Major Feedstocks

Feedstocks	Increase/Decrease in Total Consumption[1]		
	U.F. = 0%	U.F. = 50%	U.F. = 100%
Benzene	−41	−27	0
Normal- and iso-Butane	−22	19	0
Ethane	−23	−16	0
Ethylene	−3	−4	0
Methane	1	3	2
Naphtha	1	−1	0
Propane	−8	0	0
Toluene	60	43	1

[1]Total production within the network + exogenous supply.

4   Acceptance of the new styrene process at 0% and 50% U.F., results in 36% and 68% reduction in ethylbenzene consumption, respectively. Because propylene oxide is a co-product of the hydroperoxide process to styrene, the propylene oxidation route to propylene oxide has to operate at higher production level (capacity expansion indicated at 50% U.F.) to make up for the needs for propylene oxide. As a result, butadiene becomes a better feedstock than propylene oxide for production of 1,4-butanediol, and this points out further expansion of the n-butylenes-based process (oxidative dehydrogenation route) to butadiene.

5   Acceptance of the new vinyl acetate process has a great impact on production of acetic acid and acetaldehyde. Due to the better yield of this process and by-product acetic anhydride, production of acetic acid and acetaldehyde in the model drops 8% and 46%, respectively, when the industry operates at 100% U.F. Allowing the industry to operate at lower utilization factors results in more reduction in acetic acid production and finally at 0% U.F., just enough acetaldehyde is produced to satisfy its exogenous demand and acetic acid consumption drops by 37%.

In studying the impact of the new technologies on the 1985 industry, the 10 new processes were introduced into the model as a single block. This methodology is comparable to a decision-maker taking into account the simultaneous impact of all the new technologies. An alternate method is to study the attractiveness of each individual process separately. This gives an estimate of the cost improvement (reduced cost) required to make the process acceptable to the model industry.

Table 2.16 lists the reduced cost of the new technologies, assuming the 1985 industry operates at 50% of the 1977 built capacities. The negative reduced costs represent incentives for the processes to be accepted by the optimal industry. This is a measure of the profitability of the processes and represents the cost savings resulting from the operation of such technologies. All the six accepted processes listed in Table 2.16 plus the stilbene route to styrene show negative reduced costs, but the toluene route to styrene (process N8) is preferred when all the new processes are assessed as a block.

The study in this section illustrates how the model can be used in technology planning to estimate the relative attractiveness of new technologies, the cost improvement required for the new technologies to be accepted by the industry, and the potential impact on the whole industry, when the new processes are adopted.

SENSITIVITY STUDIES

Aside from variations on the utilization factor and assessment of new technologies, the industry is subject to constant perturbations from the economic environment, such as changes in supply/demand patterns and fluctuations in

Table 2.16 Cost Improvement Required for New Technologies to be Accepted in 1985 When the Industry Expands beyond 50% of the 1977 Built Capacities

Chemical	Process	Cost Improvement, ¢/lb of Product	
		Block Introduction	Individual Introduction
Aniline	Direct amination of benzene	0	−57.05
	Toluene via benzoic acid route	0	−59.77
Ethylene glycol	Synthesis gas and formaldehyde route	36.21	36.21
	Direct hydration process	35.36	35.36
Maleic anhydride	n-Butenes route	0	−9.52
Phenol	Toluene route	0	−11.28
Styrene	Dimerization of butadiene	76.83	35.62
	Toluene route	0	−44.87
	Toluene dimerization and disproportionation of stilbene	0	−43.61
Vinyl acetate	Carbonylation of acetic acid	0	−54.33

chemical prices and energy expenditure. In this section we illustrate the use of the economic model to analyze the behavior of the industry under different scenarios.

Perturbations in Energy Consumption of the Industry

The model is used to test the impact on the industry's selection of processes in 1985, assuming that the fuel oil and power costs do not go up as rapidly as gas, naphtha, and gas oil prices. This is visualized as a possible situation if distillate demands rise rapidly due to increasing needs for heating oil and diesel fuel, naphtha demand for chemical feedstock also continuing relatively strong, and gas supplies remaining tight. On the other hand, partly because of increasing use of heavier crude oil and great inroads made by coal and nuclear energy into power generation, industry could become long on heavy fuel oil by 1985 and fuel oil prices might rise more slowly than prices of the light, clean products.

To study the impact of this situation on the chemical industry, a case is constructed in which the 1985 costs for utilities are assumed to be only 50% of the estimated values. The major impact of this fuel cost adjustment on the 1985 industry's process selection is reflected in Table 2.17 and some of the more significant changes are summarized as follows:

1 Low fuel cost mainly affects production of acetic acid and acetaldehyde. As a result, the n-butane route to acetic acid becomes the most attractive pro-

cess and makes acetic acid the only feedstock for production of acetic anhydride.

2 With lower fuel oil and power costs, the industry tends to favor more butadiene production but lower acetylene. This is simply due to the attractiveness of the butadiene route to chloroprene and the model recommends a
utilization factor of 79% for this process.

Because utilities usually represent a small portion of the total processing cost,
the impact on process selection is expected to be relatively minor. The results of
the runs tend to support this general observation.

Impact of Shortages in Propane and Natural Gas Supply on the Industry

In this section the model is used to study the effects of perturbations in present
patterns of chemicals supply. Essentially all of the industry's primary feedstocks
are now derived from petroleum and natural gas. However, the possibility of
natural gas and propane shortages may cause the industry to turn away from
these chemicals as feedstock suppliers. For instance, it is suggested that synthesis gas and methane produced from coal would replace natural-gas-derived
feedstocks in some applications, and it is also predicted that there will be shortages in propane supply for chemical uses. The model is used to consider these
two scenarios.

Consider first the scenario in which propane for chemical use is in short supply for the chemical industry. Three 1985 cases are studied. Case 1 assumes the
1985 industry operates at 50% of the 1977 built capacities (50% U.F.). Case 2
uses the same bases as Case 1 except that only 50% of the previously available
propane for chemical use is accessible in 1985 and the capacity limitations of
the propane-based ethylene processes are removed to give more flexibility to the

Table 2.17 The U.S. Process Capacity Expansion Needs in 1985
Estimated Fuel Cost vs. Low Fuel Cost (U.F. = 50%)

Chemicals	Processes[1]	Capacity Expansion Needs (Millions of Pounds)	
		Estimated Fuel Cost	Low Fuel Cost
Acetaldehyde	1	2190	0
Acetic acid	4	430	410
	6	0	2585
Acetylene	16	395	310
Butadiene	41	490	580

[1]The process numbers refer to Table 2.2.

industry. Case 3 is similar to Case 2 except the exogenous supply of propane is set at zero meaning no propane will be available to the 1985 chemicals industry.

Tables 2.18 and 2.19 show the major impact of propane shortage on the 1985 industry's capacity expansion needs and process utilization factors, respectively. The changes in consumption of primary feedstocks are also shown in Table 2.20. We make the following observations:

1   The propane shortage tends to reduce the consumption of ethylene by 6%. As a result, ethylene dichloride production within the network drops more than 5% and acetylene also becomes a good feedstock for production of vinyl chloride. Therefore, acetylene from the submerged flame process becomes very attractive, partly because ethylene is its co-product.

2   Switches also occur in ethylene-making processes, and naphtha becomes an even more attractive feedstock. Industry also has to expand the pyrolysis of ethane process in order to keep up with the needs for ethylene.

3   Acetic acid becomes the superior feedstock for acetic anhydride production. As a result, the *n*-butane route to acetic acid is picked by the model for expansion which results in a 62% increase in the total consumption of *n*-butane.

4   The attractiveness of propylene based acrylonitrile reduces the consumption of hydrogen cyanide in the model.

Impact of Coal Gasification

Consider the next scenario in which shortages may occur in natural gas as a feedstock supplier. In order to close up the coal cycle in the technology catalog, three processes are also added based on synthesis gas conversion to pure carbon monoxide and hydrogen. The 1985 cases studied here are identical to the propane cases except in Case 2. Instead of the propane-based ethylene processes, capacity limitations of the natural gas and/or methane-based processes for acetylene, ammonia, carbon monoxide, hydrogen, hydrogen cyanide, methanol, and synthesis gas are removed.

The major impact of natural gas shortage on the industry's process selection and raw material consumption are shown in Tables 2.21, 2.22, and 2.23 and summarized as follows:

1   The deficiency of natural gas makes naphtha the better feedstock for ammonia production. As a result, methane-based ammonia is forced out of the optimal industry in Case 3.

2   The attractiveness of acetylene from the submerged flame process makes this chemical a good feedstock for vinyl chloride production. Therefore, production of ethylene dichloride drops more than 5% which in turn reduces the production of ethylene-based vinyl chloride.

Table 2.18 Impact of Propane Shortage on the Industry's Capacity Expansion Needs in 1985

Chemicals	Processes[1]	Capacity Expansion Needs (Millions of Pounds)		
		Case 1	Case 2	Case 3
Acetaldehyde	1	2,190	0	0
Acetic acid	6	0	2,560	2,560
Acetylene	12	10	500	500
	16	395	0	0
Ethylene	83	0	460	3,960
	82 + 85	16,150	20,200	22,000
Hydrogen cyanide	108	670	115	115
Vinyl chloride	175 + 176	2,100	1,930	1,930
	177	0	10	10

[1] The process numbers refer to Table 2.2.

Table 2.19 Impact of Propane Shortage on the Capacity Utilization Factors in 1985

Chemicals	Processes[1]	Capacity Utilization Factors		
		Case 1	Case 2	Case 3
Acetaldehyde	1	—[2]	100	100
Acetic acid	6	67	—[2]	—[2]
Acetylene	16	—[2]	50[3]	50[3]
Acrylonitrile	23	50[3]	100	100
Ethylene	80	50[3]	0	0
	84	50[3]	30	3
Ethylene dichloride	90 + 91	20	19	19
Vinyl chloride	177	50[3]	—[2]	—[2]

[1] The process numbers refer to Table 2.2.
[2] Expansion needed.
[3] Forced to the industry.

3 Processes based on coal become very competitive resulting in a switch to synthesis-gas-based methanol. As a result, n-butane becomes better feedstock than methanol for acetic acid production and this results in a 14% increase in the total consumption of n-butane.

4 The attractiveness of propylene-based acrylonitrile reduces the consumption of hydrogen cyanide in the model.

Table 2.20 Impact of Propane
Shortage on the Total Consumption
of Primary Feedstocks in 1985

| Chemicals | Increase/Decrease in Total Consumption[1] (%) | |
	Case 2[2]	Case 3[2]
n-Butane	62	62
Ethane	−53	12
Ethylene	−6	−6
Methane	−2	−2
Naphtha	21	31
Propane	−48	−95

[1]Total production within the network + exogenous supply.
[2]Using Case 1 as the basis.

Table 2.21 Impact of Natural Gas Shortage on the Industry's Capacity
Expansion Needs in 1985

| Chemicals | Processes[1] | Capacity Expansion Needs (Millions of Pounds) | | |
		Case 1	Case 2	Case 3
Acetaldehyde	1	2,190	2,225	2,225
Acetic acid	4	430	0	0
	6	0	275	275
Acetylene	12	10	655	700
	16	395	0	0
Ethylene	82 + 85	16,150	16,100	17,500
Hydrogen cyanide	108	670	670	225
Vinyl chloride	175 + 176	2,100	2,100	1,980

[1]The process numbers refer to Table 2.2.

Although these two scenarios are extreme cases, they serve to demonstrate the
applicability of the model to problems of long-range industrial development.

CONCLUDING REMARKS

The economic model presents an ideal industry situation. However, with careful
use of constraints and bounds, the model provides a reasonable representation

Table 2.22 Impact of Natural Gas Shortage on the Capacity Utilization Factors in 1985

Chemicals	Processes[1]	Capacity Utilization Factors (%)		
		Case 1	Case 2	Case 3
Acetic acid	4	$-^2$	50^3	50^3
	6	67	$-^2$	$-^2$
Acetylene	14	50^3	0	0
	16	$-^2$	50^3	50^3
Acrylonitrile	23	50^3	50^3	89
Ammonia	32	80	53	0
Ethylene dichloride	90 + 91	20	20	19
Methanol	127	93	0	0
Vinyl chloride	177	50^3	50^3	86

[1]The process numbers refer to Table 2.2.
[2]Expansion needed.
[3]Forced to the industry.

Table 2.23 Impact of Natural Gas Shortage on the Total Consumption of Primary Feedstocks in 1985

Chemicals	Increase/Decrease in Total Consumption[1] (%)	
	Case 2^2	Case 3^2
n-Butane	14	14
Ethane	-18	-18
Ethylene	0	-2
Methane	-55	-98
Naphtha	20	65

[1]Total production within the network + exogenous supply.
[2]Using Case 1 as the basis.

of the total industry. Such a model can then be used to project future industrial development for a variety of economic and technological scenarios.

The model is a useful tool for long-range planning studies, such as testing relative attractiveness of new technologies and their impact on processes and feedstocks, testing impacts of changes in feed supply, product demand and process economics on process selections, and finally studying the impact of economic growth, market variations, and price fluctuations on the chemicals industry.

This model is based on the assumption that the reported and projected prices of chemicals used within the industry are reliable indicators of economic value. The integrated model of the industry discussed in the next chapter removes this assumption by focusing attention only on the value of chemicals that enter and leave the industry, and by assuming that the intermediate chemicals within the industry have value only as building blocks for end chemicals or as fuels for the process energy demands. This model is apt to be a more accurate indicator of the large changes within the industry that would disrupt pricing structures of intermediate chemicals.

CHAPTER THREE

Integrated Model of the Intermediate U.S. Chemical Industry

The economic model of the industry examined in Chapter 2 assigns a market value to intermediate chemicals. Here we assume that the value of chemicals derives from their uses in the economy at large and that prices are known only for the chemicals that enter and leave the intermediate chemicals industry. The chemicals both produced and consumed within the industry are assigned prices only when used as fuels.

In this chapter, the economic model of Chapter 2 is changed to reflect this integrated perception of the industry. The similarity of the 1985 industry as projected by the economic model and this integrated model suggests that the integrated model is a useful planning tool in the absence of reliable data on intermediate chemical prices. A case study is also presented to show the potential uses of this integrated model for the problems of long-range planning such as the attractiveness of biomass alcohol as an alternative feedstock. How alcohol produced by fermentation from biomass can be a chemical feedstock to replace petrochemical alcohol is shown. The integrated model is a useful tool for determining the price subsidy necessary to make biomass alcohol competitive. Further, the success of the integrated model in duplicating the results of the economic model for the U.S. industry increases confidence in its use in development planning.

MODEL DESCRIPTION

In the chemical industry, market prices of the intermediate chemicals are in fact one of the major deciding factors in process selection. To reflect this interaction between the industry and the economy, the intermediate chemical prices were incorporated into the model as part of the industry's total production cost (Chapter 2). But if the industry is to be built all over again, the transformations would start with the primary raw materials at one side and carried out to the end products that have direct use in the consumer market. The intermediate chemicals would be used captively within the industry and they would have no value as far as the consumer market is concerned. This totally integrated model is a rather idealized situation for the petrochemical industry in the United States, but has practical uses in developing countries.

The integrated industry's total production cost is formulated as follows:

$$\text{total production cost} = \sum_{i=1}^{N} P_i F_i + \sum_{j=1}^{M} C_j^* X_j - \sum_{i=1}^{N} H_i (Q_i - D_i)$$

The terminology is the same as in Chapter 2 except for the term C_j^* which represents the utilities costs, investments and labor-related costs, and the fixed costs associated with the raw materials portion of the process such as the cost of catalysts, additives, and fillers. Because no intermediate chemical prices are assigned and therefore no by-product credits are given, the surplus materials are only credited as heating values.

In the following sections, the potential uses of such a model in understanding the future trends of the petrochemical industry are illustrated.

PLANNING THE 1985 INDUSTRY

Under the economic environment defined in Chapter 2, the 1985 capacity expansion needs indicated by the integrated model are summarized in Table 3.1, and Table 3.2 lists the utilization factors of the processes recommended by the model.

Table 3.1 The U.S. Process Capacity Expansion Needs in 1985 When the Integrated Industry Expands beyond 0% and 50% of the 1977 Built Capacities

Chemicals	Processes[1]	Capacity Expansion Needs (Millions of Pounds) U.F. = 0%	U.F. = 50%
Acetaldehyde	1	0	30
Acetic acid	4	3,100	2,290
Acetic anhydride	8 + 9	480	480
Acetylene	16	1,080	400
Acrylamide	18	55	30
Acrylic Acid	22	110	0
Adipic Acid	25	1,660	1,280
Aniline	34	140	140
Bisphenol-A	39	465	465

Table 3.1 (*Continued*)

Chemicals	Processes[1]	Capacity Expansion Needs (Millions of Pounds)	
		U.F. = 0%	U.F. = 50%
Butadiene	41	130	490
	42	15	0
Chlorobenzene	60 + 61	30	30
Cyclohexane	65	0	335
Cyclohexanol	66 + 67 + 68	0	195
Cyclohexanone	70	365	0
Dimethyl terephthalate	72	3,295	1,345
Dinitrotoluene	73	35	35
Epichlorohydrin	74	150	105
Ethylbenzene	79	560	600
Ethylene	81 + 86 + 87	1,990	5,750
	84	13,650	5,000
Ethylene dichloride	90 + 91	3,250	2,990
Ethylene glycol	92	640	640
Ethylene oxide	95	4,300	3,250
2-Ethylhexanol	97	245	245
Glycerine	99 + 100	110	70
Hexamethylenediamine	104	570	390
Hydrogen cyanide	108	1,110	670
Hydrogen peroxide	109	0	100
	110	370	100
Isoprene	118	125	110
Maleic anydride	122	410	210
Methanol	127	320	145
Methyl methacrylate	136	1,300	900
Phenol	143	1,795	1,430
Phosgene	144	570	570
Phthalic anhydride	146	480	480
Propylene, polymer grade	148	1,235	750
	149	1,200	1,710
Propylene glycol	150	55	55
Styrene	154	6,720	3,075
Terephthalic acid, crude	166	5,935	4,220
Toluene diamine	168	20	20
Toluene diisocyanante	169	25	25
Urea	170	17,800	12,650
Vinyl acetate	172 + 173	590	590
Vinyl chloride	175 + 176	2,495	2,350
p-Xylene	182	3,290	2,350

[1]The numbers refer to process numbers in Table 2.2.

Table 3.2 Recommended Utilization Factors of the Processes in 1985 When the Integrated Industry Expands beyond 0 and 50% of the 1977 Built Capacities

Chemicals	Processes[1]	Utilization Factors	
		U.F. = 0%	U.F. = 50%
Acetaldehyde	1	97	—[2]
Acrolein	17	38	61
Acrylic acid	22	—[2]	50[3]
Ammonia	32	80	80
Benzene	37	42	62
Butadiene	42	—[2]	50[3]
s-Butanol	50	4	52
Caprolactam	53 + 55 + 56	0	88
Chlorine	59	55	61
Cumene	64	33	50[3]
Cyclohexane	65	85	—[2]
Cyclohexanol	66 + 67 + 68	9	—[2]
Cyclohexanone	70	—[2]	50[3]
Ethanol	75	77	79
Ethylene	82 + 85	0	0
Formaldehyde	98	83	88
Hydrogen Peroxide	109	0	—[2]
Isopropanol	119 + 120	94	91
Methyl isobutyl ketone	134	48	50[3]
Phenol	140	37	57
Phthalic anhydride	145	63	63
Styrene	153	17	62
Terephthalic acid, fiber grade	164	0	0

[1]The numbers refer to process numbers in Table 2.2.
[2]Expansion needed.
[3]Forced to the industry.

The following observations are made by analyzing the behavior of the industry under different utilization factors.

1   In the integrated industry, low-pressure carbonylation of methanol and oxidation of p-xylene to terephthalic acid (crude) are the most attractive routes to produce acetic acid as a major and/or side-product. But when the industry operates at 50% U.F., acetaldehyde and also n-butane-based acetic acid become responsible for 23% of this chemical's total production within the network. By relaxing the capacity limitations (U.F. = 0%), acetaldehyde and n-butane routes to acetic acid are replaced by the alternate technologies. As a result, acetaldehyde's total consumption drops by more than 47%.

2 Based on the economic environment defined to the model, isobutylene-based methyl methacrylate, direct sulfonation of benzene to phenol, and also autoxidation of isopropanol to hydrogen peroxide (resulting in acetone as a side-product) are among the processes which are favored over the alternate routes in the model. As a result, by allowing the industry to reach its optimal structure with no capacity limitations, the total consumption of acetone and cumene drop 14% and 35%, respectively.

3 In the integrated model, acetylene is the preferred feedstock for production of acrylic acid, chloroprene, and ethyl acrylate. Therefore, when the utilization factor is reduced to zero, the acetylene-based processes for these chemicals replace the alternate technologies.

4 Ethylene and butadiene are also preferred feedstocks for acrylonitrile and hexamethylenediamine, respectively. As a result, at 0% U.F., cyanation/oxidation of ethylene and hydrocyanation of butadiene are favored over the competing processes, which in turn results in further expansion of the hydrogen cyanide-making process.

A comparison of the processes dominant in the actual industry and those in the integrated industry shows that the inclusion of intermediate chemicals' prices does not have a significant impact on the optimal structure of the industry. Figure 3.1 represents the processes selected by both models in 1985 assuming that the actual and integrated industries operate at 0% of the 1977 built capacities. It is evident that only 16 of the processes that appear in the optimal structure of both industries differ from each other and a process agreement of 81% is obtained.

SENSITIVITY STUDIES ON FERMENTATION ALCOHOL

There is continuing evidence that ethanol may become increasingly important both as a fuel and as a source of chemicals. It is recommended that fermentation alcohol be regarded as a permanent alternative source of fuel and chemicals, particularly in countries lacking oil resources.

Brazil has one of the most ambitious alcohol programs in the world. In 1977, biomass accounted for about 28% of Brazil's total energy consumption. Wood, sugar cane, charcoal, and fermentation alcohol are the traditional biomass fuels in Brazil. Motivated by the government, the departure from an oil- to ethanol-based chemical industry presents a significant challenge to Brazil's economic, social, and political activities. Important social benefits from this program are the creation of jobs at the farm level, improvement in income distribution, and a more balanced development throughout the country. A recent report indicates that with the Brazilian government's price subsidy for ethanol some of the primary derivatives of ethanol, such as acetaldehyde and ethylene, can compete favorably with products made from petrochemical feedstocks.

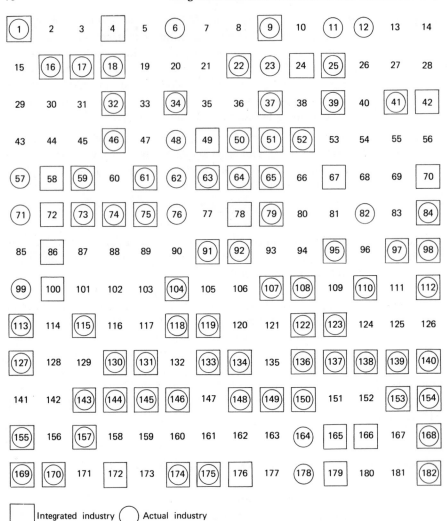

Integrated industry ☐   Actual industry ○

Figure 3.1 A comparison of the processes dominant in the actual industry and those in the integrated industry in 1985. (The numbers refer to process numbers in Table 2.2 of Chapter 2.) (U.F. = 0%.)

In general, economics favors an oil-based chemical industry. However, much can be done to improve the economics of producing fermentable materials and to advance the technology of fermentation so as to lower the ethanol output price.

In this section, the competitiveness of fermentation ethanol with petroleum feedstocks is studied under different economic environments, using the model to determine the price subsidy for ethanol. To this effect, a few 1985 cases are constructed. Case 1 represents a situation in which the industry is assumed to operate at 0% of the 1977 built capacities and only a limited supply of ethanol is available to the industry. (This case was analyzed in the previous section). The other cases use the same bases as Case 1 except it is also assumed that an unlimited supply of fermentation ethanol is available to the chemicals industry at a fraction α of the projected price of ethanol used for industrial purposes in 1985. The factor α is varied between 100 and 0% to determine the break-even point of the industry.

Tables 3.2 and 3.3 show the major impact of the ethanol price subsidy on the 1985 industry's capacity expansion needs and process utilization factors, respectively. The changes in consumption of primary feedstocks are also shown in Table 3.4. Based on the model run results, we make the following observations:

1 Supplying the industry with unlimited amounts of fermentation ethanol at the industrial ethanol price ($\alpha = 100\%$) affects the consumption/production pattern of ethylene. In other words, ethylene is no longer used for production of ethanol and this results in a 3% reduction in the total consumption of ethylene. In feedstock selection for the steam cracking processes, gas oil and propane are still preferred raw materials over naphtha and/or ethane. But gas oil becomes more attractive than before, resulting in an 8% increase in the total consumption of this feedstock. As a result, propane's contribution in ethylene production drops by 5%.

2 The structure of the optimal industry remains unchanged for α values in the range of 100–50%. By allowing the industry to operate at $\alpha = 40\%$, gas oil becomes an even more attractive feedstock for manufacture of ethylene, and ethanol becomes responsible for acetaldehyde's production within the network. In other words, oxidation of ethylene to acetaldehyde is forced out of the optimal industry in favor of the ethanol route.

The switch from ethylene to ethanol for the manufacture of acetaldehyde was actually expected and results of the runs tend to support this general observation. However, supplying the industry with even cheaper ethanol results in a few unexpected changes in the optimal structure of the industry. These changes are summarized as follows:

1 At $\alpha = 35\%$, acetaldehyde becomes an attractive feedstock for the production of crude terephthalic acid (TPAc); the resulting production of acetic acid as a by-product in turn forces the methanol route (to acetic acid) to operate at a lower production level. Because methyl ethyl ketone (MEK)-based TPAc is no longer attractive, the total consumption of MEK drops by 53% freeing n-butenes and therefore changing the technology selection for butadiene. In other words, to prevent having surplus n-butenes, n-butane-based butadiene is forced out of the optimal industry and n-butenes become the only feedstock for production of butadiene.

Table 3.3 Impact of Ethanol Price Subsidy on the Industry's Capacity Expansion Needs in 1985.

Chemicals	Processes[1]	Capacity Expansion Needs (Millions of Pounds)					
		Case 1	α = 100-50%	α = 40%	α = 35%	α = 30%	α = 20%
Acetaldehyde	3	0	0	450	3,710	5,640	5,900
Acetic acid	4	3,100	3,100	3,100	1,600	0	0
Butadiene	41	130	130	130	1,000	1,000	1,000
	42	15	15	15	0	0	0
Ethylene	81 + 86 + 87	1,990	2,470	2,615	2,615	2,615	11,165
	84	13,650	11,950	11,440	11,440	11,440	0
Methanol	127	320	320	320	0	0	0
Terephthalic acid, crude	166	5,935	5,935	5,935	765	35	0
p-Xylene	182	3,290	3,290	3,290	3,500	3,530	3,550

[1]The numbers refer to process numbers in Table 2.2.

Table 3.4 Impact of Ethanol Price Subsidy on the Industry's Capacity Utilization Factors in 1985.

Chemicals	Processes[1]	Capacity Utilization Factors (%)					
		Case 1	$\alpha = 100-50\%$	$\alpha = 40\%$	$\alpha = 35\%$	$\alpha = 30\%$	$\alpha = 20\%$
Acetaldehyde	1	97	97	0	0	0	0
	3	0	0	—[2]	—[2]	—[2]	—[2]
Acetic acid	4	—[2]	—[2]	—[2]	—[2]	0	0
Butadiene	42	—[2]	—[2]	—[2]	0	0	0
Ethanol	75	77	0	0	0	0	0
Ethylene	84	—[2]	—[2]	—[2]	—[2]	—[2]	3
Methanol	127	—[2]	—[2]	95	95	84	84
Terephthalic acid, crude	166	—[2]	—[2]	—[2]	—[2]	—[2]	0

[1] The numbers refer to process numbers in Table 2.2.
[2] Expansion needed.

Table 3.5 Impact of Ethanol Price Subsidy on the Total Consumption of Primary Feedstocks in 1985.

Chemicals	Increase/Decrease in Total Consumption[1,2] (%)				
	$\alpha = 100-50\%$	$\alpha = 40\%$	$\alpha = 35\%$	$\alpha = 30\%$	$\alpha = 20\%$
n-Butane	0	0	−44	−44	−44
Butenes, mixed	0	0	0	0	20
Ethanol	0	38	268	405	2662
Gas oil	8	10	10	10	151
Methane	0	0	−2	−5	−5
Naphtha	0	0	−7	−21	−21
Propane	−5	−7	−7	−7	−98
Xylenes	0	0	2	3	3

[1]Total production within the network + exogenous supply.
[2]Using Case 1 as the basis.

2   When α is reduced further, ethanol-based acetaldehyde becomes a domi-
    nant feedstock for the production of acetic anhydride. At $\alpha = 20\%$, ethanol
    becomes one of the major feedstocks for ethylene production, thereby forc-
    ing propane-based ethylene to operate at only 3% of its 1977 built capacity.
    However, this is a rather extreme point for the chemical industry because at
    lower values of α (less than 18%) ethanol as a chemical feedstock becomes
    cheaper than ethanol as fuel and the problem becomes economically infeasi-
    ble.

A subsidized ethanol price of less than 12.8¢/lb ($\alpha = 40\%$) represents a rather
unrealistic situation, especially at $\alpha = 20\%$, when the industry requires about
47 billion pounds of fermentation ethanol to reach its optimal structure.
However, with the economic environment given to the model, the departure
from an oil- to ethanol-based chemical industry would occur at a subsidized
ethanol price of 12.8–16¢/lb ($\alpha = 40$–50%) in 1985.

Recent studies have been made of the fermentation products: ethanol,
isopropanol, n-butanol and 2.3-butanediol (see "Economic Evaluation of the
Large-Scale use of Biomass as a Source of Feedstocks for the U.S. Chemical Industry" B. Palsson. S. Fathi-Afshar, D. F. Rudd, and E. N. Lightfoot, *Science*,
1981).

CONCLUDING REMARKS

The economic model of the industry uses estimates of the market prices of intermediate chemicals and as a result must be considered as valid only for per-

turbations away from a stable industry. The integrated model represents an ideal situation in which no chemical value is assigned to intermediate chemicals and they become valuable as they are integrated into the chains of processes leading to valuable final products. The fact that both these models largely encompass the same projected industry indicates that the current industry in the United States has been well integrated through the forces of the market place.

CHAPTER FOUR

Systems Study of Interchangeable Chemical Products

Chemical products are useful for the service provided the economy in supplying fibers, plastics, rubber, and other synthetic materials. There are many similar plastics, fibers, rubbers, and other materials that can be used interchangeably within certain limits. A study of the intermediate chemicals industry without consideration of the outlet of its products in the economy is necessarily incomplete. We now extend our study to chemical products in the U.S. economy and introduce the idea of many chemical products performing aggregate functions in the economy.

In this chapter, the systems analysis of intermediate chemicals is extended to include aggregate end products such as plastics and resins, man-made fibers, and synthetic rubbers. An analysis of the 1977 and 1985 U.S. industry is discussed including the introduction of new polymer technologies. We also illustrate the use of the model in assessing the exchange of thermoplastics in a pipe and tubing functional aggregate.

FUNCTIONAL AGGREGATION

The chemical and allied products industry plays an essential role in the economy today. Nearly 400 companies in the United States are involved in the transformation of basic chemicals into products such as plastics and resins, textile fibers, protective coatings, pharmaceuticals, soaps and detergents, personal care products, fertilizers, pesticides, and even food for the consumer market. Economic studies can forecast changes in the demand for such use areas, but the forecast is more difficult for the particular cast of chemical building blocks needed to serve these functions in the economy.

The economy does not require a specific molecule such as acrylonitrile, styrene, or caprolactam. Rather the economy requires the function that such a molecule can perform, such as the building of fibers, elastomers, plastics, and solvents. Typically, similar products can be made from a wide choice of chemical building-block molecules and the current demand for a particular molecule is the result of a combination of price, availability, and function in the economy.

As an example, the total need for antifreeze in the United States for the past few decades is shown in Figure 4.1. This curve parallels the growth of the use of the internal combustion engine in automotive industrial use. Also shown is the cast of molecules used to perform the antifreeze function. Ethylene glycol, methanol, and ethyl alcohol have shared this responsibility. It is indicated that ethylene glycol was substituted totally for ethyl alcohol and methanol by 1970.

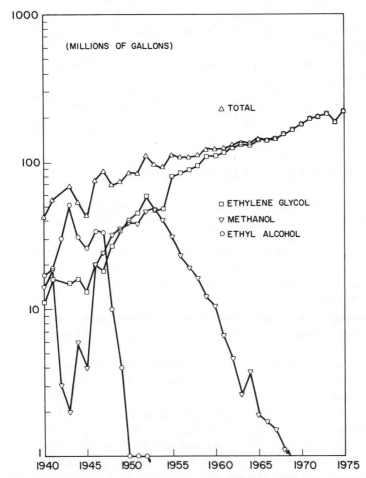

Figure 4.1   The substitution of ethylene glycol for ethyl alcohol and methanol within automotive antifreeze functional aggregate.

Another example is the consumption of barrier resins in bottles and food containers. These materials have recently been examined by the U.S. Food and Drug Administration (FDA). The major area of concern is acrylonitrile and its possible extraction into the foodstuff, and hydrocyanic acid, which may be generated by dehydration of the resin during processing, subsequently contaminating the foodstuff.

What would be the impact on the chemical industry if the use of acrylonitrile-based bottles were banned by the FDA? Obviously, the other chemicals such as polyethylene, polypropylene, polystyrene, polyvinyl chloride, polyethylene terephthalate (PET) resins, or even glass may carry out the responsibility of satisfying the demand for bottles and containers. As a result, this might have significant impact on the specific segments of the petrochemical industry.

Analyzing the functional demands of the chemical products points out the necessity of including the capability to determine both the technology and the products that best meet the needs of the economy. To determine the allowable substitutions that can be made among the chemical products, all the major functions that the industry provides to the economy should be identified. This information can then be expressed as constraints in the linear program and this new model can be linked to the existing industry model.

In the remainder of this chapter we attempt to illustrate the substitutions that have been made among the chemical products within the major segments of the industry such as plastics and resins, man-made fibers, and synthetic rubbers.

PLASTICS MATERIALS AND SYNTHETIC RESINS

Plastics materials and synthetic resins make up one of the newer, but largest and most dynamic parts of the chemicals industry. Starting with the billiard ball and the celluloid collar before the turn of the century, these materials have mushroomed since the mid-1940s into a world-wide, multibillion-dollar industry. Companies in the United States alone produced more than 21 billion pounds of plastics and resins in 1971 with a merchant sales value of 3.7 billion dollars. Between 1948 and 1972 production of plastics and resins increased tremendously from 2.1 billion pounds to 25.4 billion pounds for an average annual growth rate of 10.9%. In the more recent 1963–1972 period the growth rate was 11.7%.

A few basic characteristics account largely for this phenomenal growth. The raw materials for plastics, derived mainly from petroleum, are generally cheaper than competing materials. Plastics and resins of widely different properties can be tailored to the requirements of the end use and they often can be fabricated more cheaply than competing materials. These properties have led to revolutionary changes in the paint and coating industry, the packaging industry, the houseware industry, and others. Figure 4.2 shows the basic structure of the plastics and resins industries.

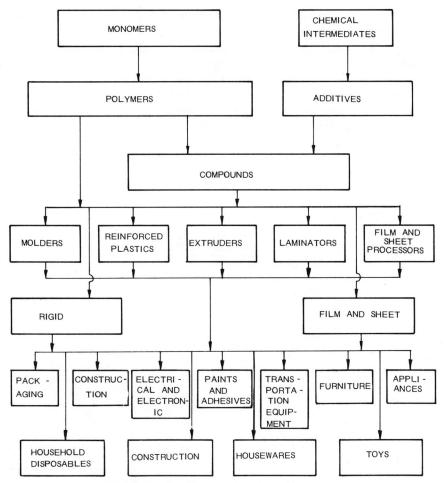

Figure 4.2 Structure of the plastics industries.

Synthetic resins are generally defined as long-chain polymers or mac-romolecules of high molecular weight. In polymerization, the macromolecule is built up by addition of multiples of the low-molecular-weight unit termed "monomer." The terms "resin" and "plastic" are used interchangeably, but not all resins can be used as plastics materials. Strictly speaking, a plastic is a relatively tough resin with a molecular weight between 10,000 and 1,000,000 that can be formed into solid shapes. However, in common industrial ter-minology a fabricated polymer is generally called a plastic and an unfabricated polymer, a resin. The dispersions of resins in water are known as latexes.

Plastics and resins have great versatility because of the variety of chemicals used in their manufacture. There are about 35 different basic polymers and thousands of different types of compounds used in plastics. The five major types, however, are polyethylene, vinyls, styrenes, phenolics, and polypropylene. Table 4.1 shows the major plastics and resins materials with the intermediate chemicals needed for their production.

All plastics are classified as either thermoplastics or thermosets. Thermoplastics soften with application of heat and regain their hardness on cooling. Thermosetting materials set into permanent shape with initial treatment of heat and pressure and do not soften on reheating.

Figures 4.3 to 4.6 show production/consumption of the major thermoplastics and thermosets by resin type in the United States for the 1940–1978 period. As

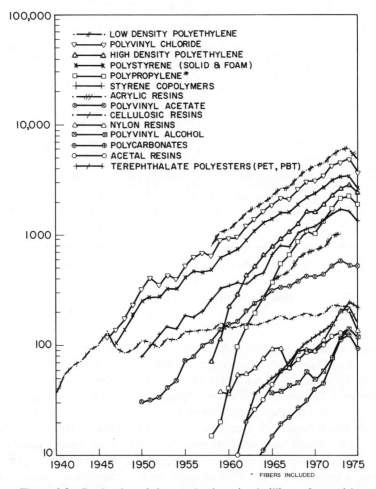

Figure 4.3 Production of thermoplastic resins (millions of pounds).

Table 4.1 Chemical Building Blocks of the Major Plastic Materials and Synthetic Resins

Plastics and Resins	Required Molecules
ABS	Acrylonitrile, butadiene, styrene
Acetal resins	Formaldehyde
Cellulosic esters	Cellulose (wood pulp and cotton linters), acetic acid, acetic anhydride
Epoxy resins	Epichlorohydrin, bisphenol-A
Melamine-formaldehyde	Melamine, formaldehyde
Nylon 6	Caprolactam
Nylon 66	Adipic acid, hexamethylenediamine
Phenol-formaldehyde	Phenol, formaldehyde
Polyacrylates	Methyl methacrylate, ethyl acrylate
Polycarbonate	Bisphenol-A, phosgene
Polyether Polyols	Propylene glycol, ethylene glycol, propylene oxide, ethylene oxide, glycerine, sorbitol, phosphoric acid
Polyethylene	Ethylene
Polyethylene terephthalate	Dimethyl terephthalate, terephthalic acid, ethylene glycol
Polypropylene	Propylene
Polystyrene	Styrene
Polyurethanes	Toluene diisocyanate, polyether polyols, methylene diphenylene diisocyanate
Polyvinyl acetate/polyvinyl alcohol	Vinyl acetate
Polyvinyl chloride	Vinyl chloride
SAN	Styrene, acrylonitrile
Urea-formaldehyde	Urea, formaldehyde
Unsaturated polyesters	Maleic anhydride, phthalic anhydride, isophthalic acid, propylene glycol, ethylene glycol, propylene oxide, styrene

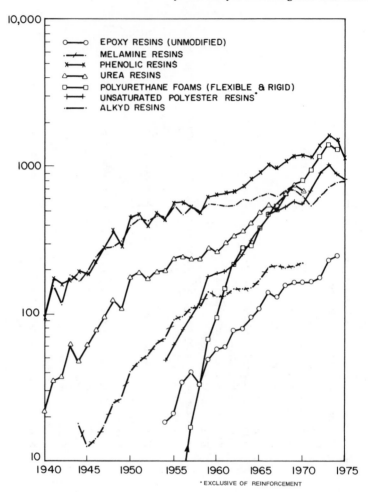

Figure 4.4   Production of thermosetting resins (millions of pounds).

shown, polyethylene is the leading plastic material with a consumption of about 9.9 billion pounds in 1978, followed by vinyls, styrenes, and polypropylene. Much of the growth of plastic materials has come from the replacement of natural materials such as wood, glass, paper, and metal in many different segments of the economy.

End-use analysis, however, shows that substitutions occur even among synthetic materials. For example, a comparative study of different plastic materials, show the potential future interaction between polypropylene and

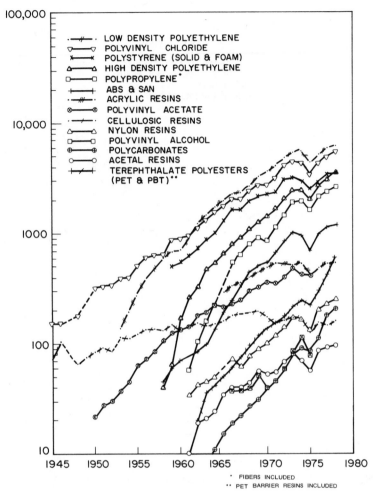

Figure 4.5 U.S. consumption of thermoplastic resins (millions of pounds).

high-density polyethylene in the injection molding area. On the basis of feedstock availability and costs, mechanical properties, and technological improvements, there will be incentives for part of the high-density polyethylene injection molding users (e.g., automotives, appliances, housewares, toys, medical, and furniture) in the United States to switch to polypropylene in the more vulnerable areas such as housewares, toys, caps and closures, and furniture. Studies of similar kind also show that polypropylene is competing with polyvinyl chloride, polystyrene, ABS, and high- and low-density polyethylene in

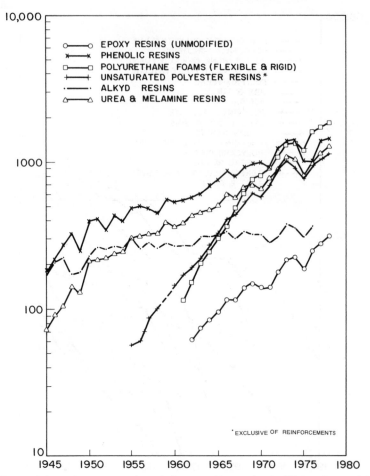

Figure 4.6 U.S. consumption of thermosetting resins (millions of pounds).

areas such as bottles, pipe and tubing, and film. Figures 4.7 to 4.10 show the consumption of synthetic resins in wire and cable, pipe and tubing, bottles, and film and sheet for the 1960–1978 period. Nearly 200 million pounds of polyvinyl chloride film were used in the food-packaging industry in 1978. Another problem in substitution is arising with the discovery of the cancer-producing property of vinyl chloride. Should the use of polyvinyl chloride be banned for packaging food, the industry must respond by the increased production of the products such as polyethylene, polypropylene, or polystyrene capable of filling the role that polyvinyl chloride now plays in the packaging industry. However,

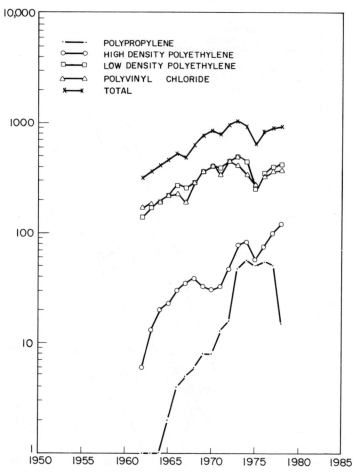

Figure 4.7 U.S. consumption of resins for wire and cable extrusion (millions of pounds).

deficiency of the basic raw materials, ever-increasing oil prices, and Environmental Protection Agency's (EPA) regulations will force the chemicals industry to suggest workable solutions and to allocate the materials in the most economical fashion to the industry as a whole.

MAN-MADE FIBERS

Fibers are fundamental units used in the fabrication of textile yarns and fabrics having a length of at least 100 times its diameter and usually having a

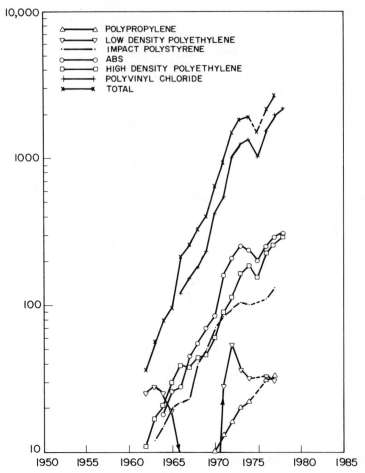

Figure 4.8 U.S. consumption of resins for pipe and tubing (millions of pounds).

definitely preferred orientation of its crystal unit cells with respect to a specific axis. Basically there are two classes of fiber: (1) natural fibers, those of animal, vegetable, or mineral origin (e.g., cotton, wool, silk); and (2) man-made fibers, those fibers not found in nature. These include cellulosics, which are partially synthetic (e.g., rayon, cellulose acetate), and wholly synthetic fibers, which are composed of materials made entirely by chemical synthesis (e.g., acrylic, nylon, polyester).

Table 4.2 shows the major fiber materials with the intermediate chemicals needed for their production. There are more than 17 man-made fibers available

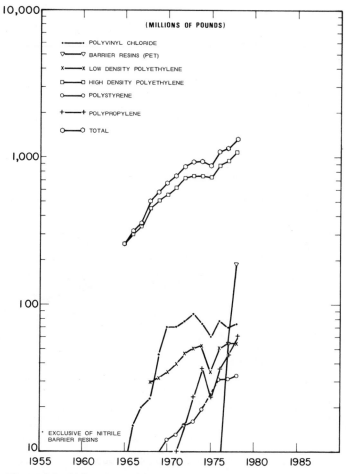

Figure 4.9 U.S. consumption of resins for blow-molded bottles.*

to the world's textile industry. That number, however, does not tell the whole story of the man-made-fiber revolution. For each of the 17 main types there are up to several thousand variations, each individually tailored to meet the requirements of a specific end use.

Man-made fibers are made by extruding filaments from a polymer melt or solution through small orifices in a spinneret. Filaments can be produced in various sizes ranging from very small diameters to large bristles. Single strands, called *monofilaments*, usually have relatively large diameters. They are most frequently used for non-textile applications such as fishing lines and lawn fur-

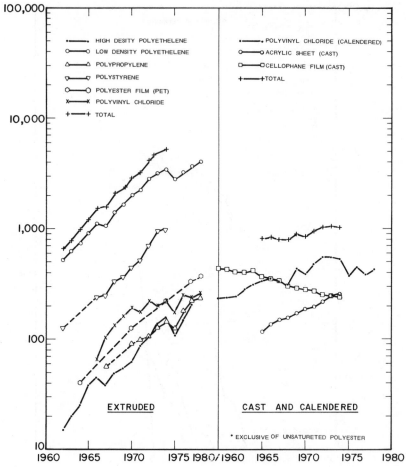

Figure 4.10   U.S. consumption of resins for film and sheet* (millions of pounds).

niture. The more conventional textile *multifilament yarns* are made up of several small diameter filaments twisted together. *Tow,* a group of filaments in the form of a rope, is cut up into uniform lengths. These fibers, or *staple,* are crimpled and twisted to approximate the properties of natural fibers and then spun into yarns.

Figures 4.11 to 4.14 show the U.S. production and consumption of different man-made fibers for the period of 1950–1978. As shown, polyester is the leading synthetic fiber with the producers' domestic shipment of about 3.5 billion pounds in 1978 in 1978 followed by nylon and olefin fibers.

Table 4.2 Chemical Building Blocks of the Major Fibers

Fibers	Required Molecules	Typical Trade Names
Cellulosic fibers		
Rayon	Wood pulp and cotton linters	Bemberg, Fibro
Acetate and Triacetate	Wood pulp and cotton linters, acetic anhydride, acetic acid	Acele, Celanese
Noncellulosic fibers		
Acrylic	Acrylonitrile	Orlon, Zefran
Aramid	m- and p-phenylenediamine, iso- and terephthaloyl chloride	Kevlar, Normex
Fluorocarbon	Tetrafluoroethylene, fluorinated ethylene-propylene copolymers	Teflon
Modacrylic	Acrylonitrile, vinyl chloride, vinylidene chloride, vinyl acetate	Verel, Dynel
Nylon	Hexamethylenediamine, adipic acid, caprolactam	Perlon (for nylon 6)
Olefin	Propylene, ethylene	Herculon
Polyester	Dimethyl terephthalate, terephthalic acid, ethylene glycol	Dacron, Kodel, Fortrel, Vycron
Saran	Vinylidene chloride, vinyl chloride	Saran
Spandex	Diisocyanates, polyester or polyether polyols, diamines	Lycra, Vyrene, Glospan
Vinyon	Vinyl chloride, vinyl acetate	
Textile glass fibers	Glass	Fiberglass
Metallic fibers	Aluminum, stainless steel, various plastics	

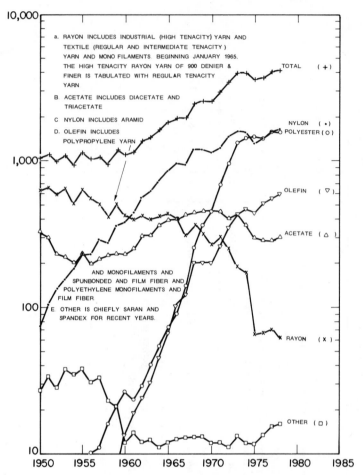

Figure 4.11 Production of man-made fibers (yarn and monofilaments) (millions of pounds).

Although natural fibers such as cotton, wool, and silk have excellent textile properties, man-made fibers as a group have greater versatility and can be tailored for improved washability, durability, tensile strength, and resistance to soiling and shrinking. These advantages have allowed man-made fibers to gradually replace cotton and wool in many different applications and virtually eliminate silk as a commercial fiber. Figure 4.15 shows the replacement of natural materials by man-made fibers over the past years, indicating the mill consumption of 3.2 billion pounds for cotton and wool in 1978 which shows an average 3.3% annual decrease since 1950.

End-use analysis of fiber materials points out the fact that other than contin-

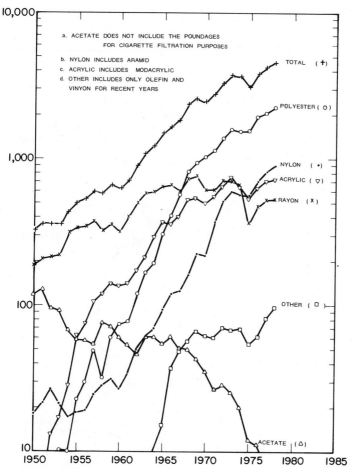

Figure 4.12 Production of man-made fibers (staple and tow) (millions of pounds).

uing replacement of natural materials by man-made fibers, interactions occur even among the synthetic fibers. This is illustrated in Figures 4.16 and 4.17 which show the consumption of fibers in tires (mainly tire cord) and carpets and rugs for the 1960–1978 period. It can be seen that noncellulosic fibers account for almost all the fiber materials used in carpets and rugs since early 1970s. Although nylons were the leading noncellulosic fibers used in tires, polyester has started to compete closely with nylons since the mid-1970s and its consumption is expected to grow at a somewhat faster rate in future years. Interactions have also occurred between polyester, acrylics, and olefins fibers in carpets and rugs, but consumption of acrylics is expected to decline and polyester and

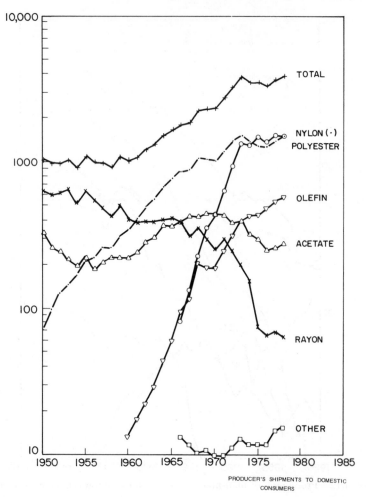

Figure 4.13   U.S. consumption* of man-made fibers (yarn and monofilaments) (millions of pounds).

olefins remain competitive as face fibers. The growth pattern of man-made fibers in different markets also indicates that polyester will probably remain very competitive with acrylics in blanketing, and pass rayon in draperies and curtains.

SYNTHETIC RUBBERS

Synthetic rubbers (elastomers) are organic polymers that are capable of quick recovery from large deformations. Three decades ago, natural rubber held

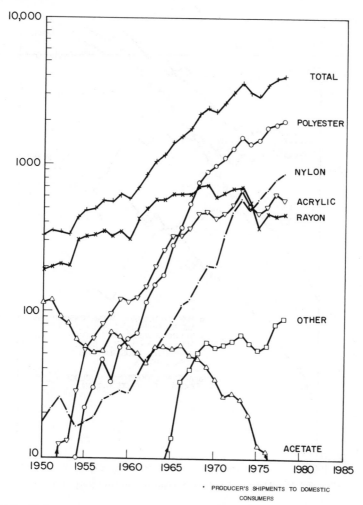

Figure 4.14 U.S. consumption* of man-made fibers (staple and tow) (millions of pounds).

99.6% of the market in the United States. Elastomers, however, penetrated the market, capturing more than half of the market by 1951. In 1977, natural rubber shared only 25% of the market and synthetics dominated with about 75%. Figure 4.18 shows the relative consumption of natural rubber versus synthetic rubbers for the 1950–1977 period.

Synthetic rubbers were first produced commercially in Germany during World War I. The U.S. industry was established on a high volume basis during World War II when the government, anticipating the loss of natural rubber sources, built enough synthetic capacity to meet wartime requirements.

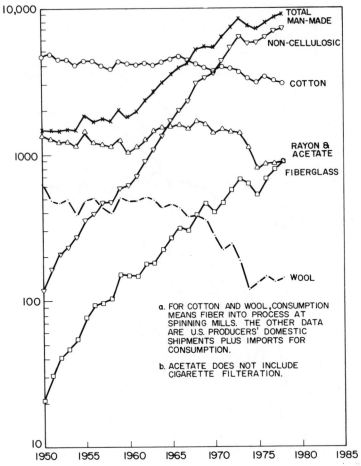

Figure 4.15   U.S. textile fiber consumption (millions of pounds).

Styrene-butadiene rubber (SBR) was picked as the general-purpose elastomer to substitute natural rubber. Now there are more than eight major elastomers available to the U.S. rubber industry. Table 4.3 lists the major synthetic rubbers produced in the United States and the chemical intermediates used in their structure. Most elastomers are compounded with various additives like carbon black and rubber-processing chemicals such as antioxidants, antiozonants, stabilizers, and accelerators to meet the requirements of a specific end use. Figures 4.19 and 4.20 show the U.S. production/consumption of natural and synthetic rubbers for the period of 1950–1978.

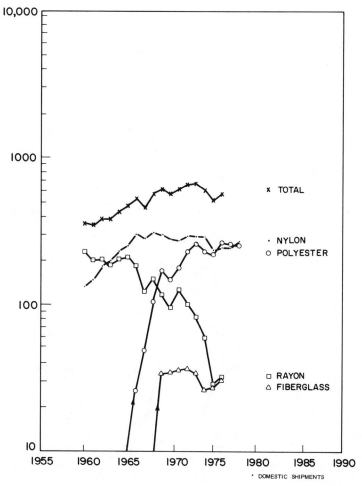

Figure 4.16 U.S. consumption* of fibers in tires (millions of pounds).

Natural and synthetic rubbers are used largely in the automotive industry, mainly for tires and related products. Nontire automotive uses such as gaskets, wire coating, radiator hoses, and battery cases account for about half of the remainder. The other end uses include toys, sporting goods, foamed products, shoes, and so on. Table 4.4 shows the end-use breakdown of the natural and synthetic rubbers in 1963 and 1971.

Styrene-butadiene rubber was the leading rubber material in the United States with a domestic consumption of about 3 billion pounds in 1978. Tires

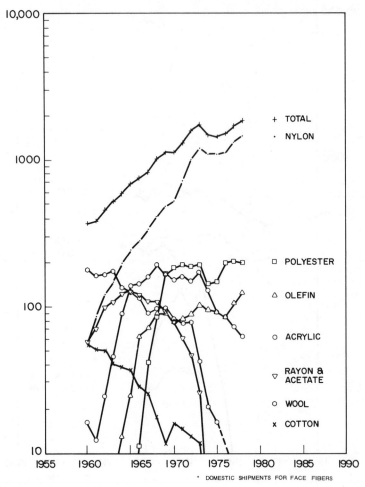

Figure 4.17 U.S. consumption of fibers in carpets and rugs* (millions of pounds).

and tire products have always made up the largest end-use market for SBR. In 1978, tires and tire products accounted for more than 66% of SBR's domestic consumption. This is shown in Figure 4.21 which represents the U.S. consumption of rubbers in tires and tire products since 1970. Superior wear characteristics of SBR make this compound more desirable than polybutadiene, natural rubber, and polyisoprene. The degree of substitution, however, is determined by the fluctuating prices for natural rubber and the properties required in the end product.

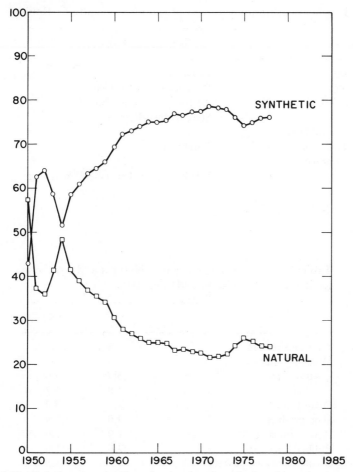

Figure 4.18   Relative consumption of natural and synthetic rubbers (percent of new).

Polybutadiene is another synthetic rubber that was expected to be used as a replacement for natural rubber, but this substance shared only 22% of the total new synthetic rubbers consumed in tires and tire products in 1975. Polybutadiene, however, is blended mainly with natural rubber in the manufacture of bus and truck tires. As in the production of passenger car tires, these formulations may vary depending on size, quality, and economics.

Cis-polyisoprene is chemically identical to natural rubber and competes with it in both tire and nontire applications. Although polyisoprene is more expensive than natural rubber, its superior physical properties make this material

Table 4.3 Chemical Building Blocks of the Major Synthetic Rubbers

Elastomers	Required Molecules
Butyl rubber	Isobutylene, isoprene
Nitrile rubber	Butadiene, acrylonitrile
Polychloroprene (neoprene)	Acetylene, chlorine, butadiene
SBR	Styrene, butadiene
Stereospecific rubbers	
Cis-polybutadiene	Butadiene
Cis-polyisoprene	Isoprene
EPDM	Ethylene, propylene, diene monomer

Table 4.4 U.S. Consumption of Natural and Synthetic Rubbers by End Use in 1963 and 1971

	1963	1971
Tires and related	61.3%	66.0%
Molded goods		
Automotive	5.7	4.6
Other	4.8	5.2
Foam rubber	3.2	3.2
Shoe products	3.0	1.9
Hose, tubing	1.9	1.9
Rubber footwear	2.0	1.6
O-rings, packing gaskets	1.5	1.5
Sponge rubber products	1.7	1.4
Solvent and latex cement	1.6	1.3
Belts, belting	1.3	1.1
Wire, cable	2.0	1.1
Coated fabrics	1.1	1.1
Floor and wall coverings	1.6	0.8
Pressure-sensitive tapes	0.6	0.5
Industrial rolls	0.4	0.5
Athletic goods	0.6	0.5
Military goods	0.6	0.5
Thread	0.5	0.5
Drugs, medical sundries	0.4	0.4
Toys, balloons	0.5	0.4
Other	3.7	4.0
Total	100.0%	100.0%

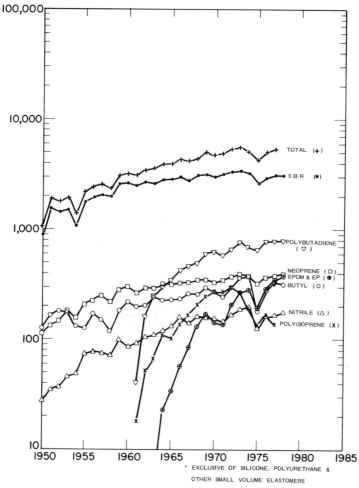

Figure 4.19 U.S. production of synthetic rubbers* (millons of pounds).

cheaper to process. Consumption of polyisoprene in tire and tire products was 85 million pounds in 1975 which represents only 3.3% of the total new synthetic rubbers used in this functional aggregate. One reason for this low contribution is the fact that polyisoprene is not used to the same extent as polybutadiene in the production of passenger car tire treads. For this application blends of SBR and polybutadiene give better results at a lower cost. The rapid growth of the production of radial tires in recent years, however, could result in a substantial increase in the consumption of polyisoprene, because this type of tire generally requires the use of polyisoprene or natural rubber.

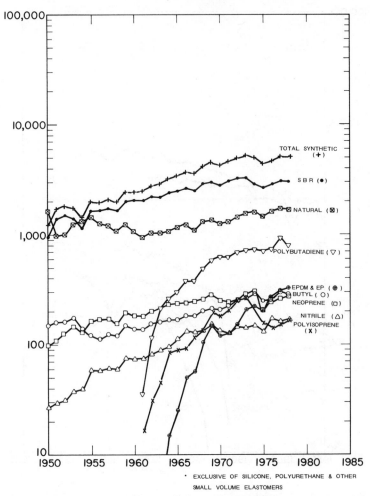

Figure 4.20   U.S. consumption of natural and synthetic rubbers* (millions of pounds).

EXTENSION OF THE MODEL TO END PRODUCTS

In the U.S. economy, intermediate chemicals rarely find direct consumer use; instead they serve as raw materials for the manufacture of end products such as plastics and resins, synthetic fibers, and synthetic rubbers. The demand for these end products can be tied to economic growth indicators. Therefore, by including the end products and their manufacturing processes, the model would be a better representation of the chemical industry. To this effect, 117 polymer

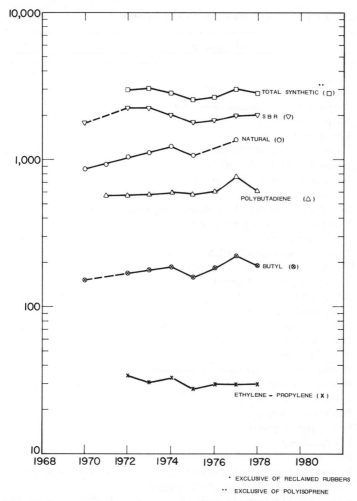

Figure 4.21 U.S. consumption of natural and synthetic rubbers* in tires and tire products (millions of pounds).

processes are added to the model which are involved in the manufacture of 72 end products. These end products are categorized in four different groups depending on the function they perform in the economy such as plastics and resins, synthetic fibers, synthetic rubbers, and thermoplastic elastomers. Tables 4.5 and 4.6 list the end products and processes added to the model. Extension of the model to end products, however, affects the exogenous supply/demand pattern of the chemicals involved in the network. Tables 4.7 and 4.8 represent

Table 4.5 End Products Included in the Model

Number	End Product
1	Acetal resins
2	Acrylic fibers
3	Acrylonitrile-butadiene-styrene (ABS)
4	Butyl rubber
5	Copolyester ethers
6	Epoxy resin (liquid DGEBA[1])
7	Epoxy resin (solid DGEBA[1])
8	Ethylene-propylene (EP) rubber
9	Ethylene-propylene-diene monomer (EPDM) rubber
10	Melamine-formaldehyde (molding compound)
11	Melamine-formaldehyde (syrup)
12	Modacrylic fibers
13	Nitrile barrier resin (BAREX)
14	Nitrile barrier resin (CYCOPAC)
15	Nitrile barrier resin (LOPAC)
16	Nitrile rubber
17	Nitrile rubber (latex)
18	Nylon 6 (chips)
19	Nylon 6 (melt)
20	Nylon 66 (chips)
21	Phenol-formaldehyde (molding compound)
22	Phenol-formaldehyde (syrup)
23	Polyacrylamide
24	Polyacrylate (latex)
25	Polyacrylate (pellets)
26	Polybutadiene
27	Polybutenes
28	Polybutylene terephthalate (PBT) (Glass-filled, self extinguishing)
29	Polybutylene terephthalate (PBT) (plain)
30	Polycarbonate (general purpose)
31	Polycarbonate (flame-resistant grade)
32	Polychloroprene (neoprene)
33	Polyether polyol (sorbitol-based hexol)
34	Polyether polyol (phosphorus-containing)
35	Polyether polyol (glycerine-based triol)
36	Polyethylene (high-density)
37	Polyethylene (low-density)
38	Polyethylene glycol
39	Polyethylene terephthalate (PET)
40	Polyethylene terephthalate (PET) barrier resin
41	Polyisobutylenes
42	Polyisoprene
43	Polyisoprene (latex)

Table 4.5 (*Continued*)

Number	End Product
44	Polymethyl methacrylate (pellets)
45	Polypropylene
46	Polypropylene glycol
47	Polystyrene (crystal grade)
48	Polystyrene (impact grade)
49	Polystyrene (expandable beads)
50	Polyurethane flexible foam
51	Polyurethane rigid foam
52	Polyvinyl acetate
53	Polyvinyl acetate (latex)
54	Polyvinyl alcohol
55	Polyvinyl butyral
56	Polyvinyl chloride (PVC) (general purpose)
57	Polyvinyl chloride (PVC) (latex)
58	Styrene-acrylonitrile (SAN)
59	Styrene-block copolymers
60	Styrene-butadiene rubber (SBR)
61	Styrene-butadiene rubber (latex)
62	Thermoplastic olefin elastomers
63	Thermoplastic polyurethanes
64	Unsaturated polyester (general purpose)
65	Unsaturated polyester (corrosion-resistant)
66	Unsaturated polyester (fire-retardant)
67	Urea-formaldehyde (molding compound)
68	Urea-formaldehyde (syrup)
69	Vinyl acetate/ethylene copolymer (latex)
70	Vinyl chloride/vinyl acetate copolymer
71	Vinylidene chloride/vinyl chloride copolymer
72	Vinylidene chloride/ethyl acrylate/methyl methacrylate terpolymer (latex)

[1]Diglycidyl ethers of bisphenol-A (DGEBA).

the updated actual/projected chemical supply/demand. Table 4.9 shows the existing process capacities for end products.

The total production cost to the industry is defined in two different ways. One is based on the production cost, defined in Chapter 2, in which market prices for intermediate chemicals are incorporated to the model as part of the manufacturing cost of the processes (Model I). The second model is based on the totally integrated industry, in which all the intermediate chemicals are

Table 4.6   Polymer Processes

Process Number	Process Description
	Plastics and resins
1	ABS graft resin by emulsion/emulsion polymerization
2	ABS graft resin by suspension/emulsion polymerization
3	ABS graft resin by bulk/suspension polymerization
4	Acetal resin from formaldehyde
5	Acetal resin from trioxane and ethylene oxide
6	Barrier resin (nitrile-BAREX) from acrylonitrile and methyl acrylate
7	Barrier resin (nitrile-CYCOPAC) from acrylonitrile and styrene (emulsion polymer)
8	Barrier resin (nitrile-LOPAC) from acrylonitrile and styrene (suspension polymer)
9	Barrier resin (PET) via terephthalic acid (continuous process)
10	Barrier resin (PET) via dimethyl terephthalate (continuous process)
11	Epoxy resin (liquid) from bisphenol-A and epichlorohydrin (batch process)
12	Epoxy resin (liquid) from bisphenol-A and epichlorohydrin (continuous process)
13	Epoxy resin (solid) from bisphenol-A and epichlorohydrin (batch process)
14	Melamine-formaldehyde (molding compound) from melamine and formaldehyde
15	Melamine-formaldehyde (syrup) from melamine and formaldehyde
16	Nylon 6 (chips) from caprolactam
17	Nylon 6 (melt) from caprolactam
18	Nylon 66 (chips) from adipic acid and hexamethylenediamine
19	Phenol-formaldehyde (Novolac molding powder)
20	Phenol-formaldehyde resol syrup (batch process)
21	Polyacrylamide (anionic) by polymerization and hydrolysis
22	Polyacrylate (47% polymer-latex) by emulsion polymerization
23	Polyacrylate (pellets) by suspension polymerization
24	Polybutenes from mixed butenes
25	Polyisobutylenes from isobutylene
26	Polybutylene terephthalate (glass-filled, self extinguishing) from DMT and 1,4-butanediol
27	Polybutylene terephthalate (plain) from DMT and 1,4-butanediol
28	Polycarbonate by continuous solution phosgenation
29	Polycarbonate by interfacial phosgenation
30	Polycarbonate by batch solution phosgenation
31	Polycarbonate (flame-resistant grade)
32	Polyether polyol (sorbitol-based hexol) from propylene oxide and sorbitol
33	Polyether polyol (phosphorus-containing) from propylene oxide and phosphoric acid
34	Polyether polyol (glycerine-based triol) from propylene oxide and glycerine
35	Polyethylene (high-density) via Phillips technology

Table 4.6 *(Continued)*

Process Number	Process Description
	Plastics and resins
36	Polyethylene (high-density) via Solvay technology
37	Polyethylene (high-density) via Union Carbide technology (gas phase process)
38	Polyethylene (high-density) via Hoechst technology
39	Polyethylene (high-density) via Montedison technology
40	Polyethylene (high-density) via Stamicarbon technology
41	Polyethylene (low density) from an autoclave reactor (compartmented)
42	Polyethylene (low density) from an autoclave reactor (backmixed)
43	Polyethylene (low density) from a tubular reactor
44	Polyethylene glycol from ethylene oxide
45	Polyethylene terephthalate (PET) from DMT and ethylene glycol
46	Polyethylene terephthalate (PET) from TPA and ethylene glycol
47	Polymethyl methacrylate (pellets) by continuous bulk polymerization
48	Polymethyl methacrylate (pellets) by batch bulk polymerization
49	Polypropylene by liquid phase bulk process (DART technology)
50	Polypropylene by vapor phase process (BASF technology)
51	Polypropylene by slurry process
52	Polypropylene by solution process
53	Polypropylene by new slurry process
54	Polypropylene glycol from propylene oxide
55	Polystyrene (crystal grade) by bulk polymerization
56	Polystyrene (impact grade) by suspension polymerization
57	Polystyrene (impact grade) by bulk-suspension polymerization
58	Polystyrene (expandable beads) by suspension polymerization
59	Polyurethane flexible foam (slab stock) from polyether polyol and toluene diisocyanate
60	Polyurethane rigid (foam sheet, fire retardant) from polyether polyols and MDI
61	Polyvinyl acetate by solution polymerization
62	Polyvinyl acetate by suspension polymerization
63	Polyvinyl acetate (62% solid-latex) by emulsion polymerization
64	Polyvinyl alcohol from vinyl acetate
65	Polyvinyl butyral by condensation of polyvinyl alcohol with butyraldehyde
66	Polyvinyl chloride (general purpose) by liquid phase bulk process
67	Polyvinyl chloride (general purpose) by suspension process
68	Polyvinyl chloride (latex) by emulsion process (continuous polymerization)
69	Polyvinyl chloride (latex) by emulsion process (batch polymerization)
70	SAN by bulk polymerization
71	SAN by continuous emulsion polymerization
72	SAN by batch emulsion polymerization
73	SAN by suspension polymerization
74	Unsaturated polyester (general purpose) from propylene glycol and anhydrides (batch fusion process)

Table 4.6 (*Continued*)

Process Number	Process Description
Plastics and resins	
75	Unsaturated polyester (general purpose) from propylene glycol and anhydrides (batch solvent process)
76	Unsaturated polyester (general purpose) from propylene oxide and anhydrides (batch process)
77	Unsaturated polyester (general purpose) by a continuous propylene oxide process
78	Unsaturated polyester (general purpose) by a continuous glycol process
79	Unsaturated polyester (corrosion-resistant) via isophthalic acid-based reaction (fusion process)
80	Unsaturated polyester (fire resistant) by fusion process
81	Urea-formaldehyde (molding compound)
82	Urea-formaldehyde (syrup)
83	Vinyl acetate/ethylene copolymer (60% solid-latex) by emulsion polymerization
84	Vinyl acetate/vinyl chloride copolymer by suspension polymerization
85	Vinylidene chloride/vinyl chloride copolymer by suspension polymerization
86	Vinylidene chloride/ethyl acrylate/methyl methacrylate terpolymer (latex) by emulsion polymerization
Synthetic fibers	
87	Acrylic fibers from acrylonitrile and methyl methacrylate (continuous solution polymerization)
88	Acrylic fibers from acrylonitrile and methyl acrylate (continuous solution polymerization)
89	Acrylic fibers from acrylonitrile and methyl acrylate (continuous suspension polymerization)
90	Acrylic fibers from acrylonitrile and vinyl acetate (batch suspension polymerization)
91	Modacrylic fibers from acrylonitrile and vinyl chloride (continuous suspension polymerization)
92	Modacrylic fibers from acrylonitrile and vinyl acetate (batch suspension polysion polymerization)
93	Modacrylic fibers from acrylonitrile and vinyl acetate (batch suspension polymerization)
94	Modacrylic fibers from acrylonitrile and vinylidene chloride (batch suspension polymerization)
Synthetic rubbers (elastomers)	
95	Butyl rubber from isobutylene
96	EPDM rubber by solution polymerization
97	EPDM rubber by suspension polymerization
98	EP rubber from ethylene and propylene
99	Nitrile rubber by cold emulsion polymerization
100	Nitrile rubber (latex) by hot emulsion polymerization

Table 4.6 (*Continued*)

Process Number	Process Description
Synthetic fibers	
101	Polybutadiene by cobalt-catalyzed polymerization
102	Polybutadiene by lithium-catalyzed polymerization
103	Polybutadiene by an iodine-ziegler catalyst
104	Polybutadiene by a nickel catalyst
105	Polychloroprene (neoprene) via butadiene
106	Polychloroprene (neoprene) via chloroprene
107	Polyisoprene by a Ziegler catalyst
108	Polyisoprene by a lithium catalyst
109	Polyisoprene (latex) from polyisoprene
110	SBR by cold emulsion polymerization
111	SBR by solution polymerization
112	SBR (latex) by hot emulsion polymerization
Thermoplastic elastomers	
113	Copolyester-ethers via DMT and 1,4-butanediol
114	Thermoplastic olefin elastomers via EPDM rubber and polypropylene
115	Polyurethane thermoplastic elastomer (ester base)
116	Polyurethane thermoplastic elastomer (ether base)
117	Styrene-block copolymers from styrene and butadiene

assumed to be used captively and, therefore, no market prices are assigned to them (Model II).

In the following sections the impact of inclusion of end products on the behavior of the U.S. petrochemical industry is studied.

Process Selection by the Model in 1977 and 1985

In order to test the reasonableness of the model with inclusion of end products versus actual industry, a study run has been made based on the economic environment defined for 1977 using the same bound structure as in "Process Selection by the Model in 1977" in Chapter 2. Part of the results are summarized in Table 4.10. It is evident from this table that the model run results in general agree with the actual industry's process selection.

Inclusion of end products, in general, has no significant impact on the intermediate chemical industry's performance. Because almost all of the polymer processes result in only one product and the alternate processes for a specific end product use basically the same raw materials.

Table 4.7 Actual/Projected U.S. Exogenous Demand[1] for End Products (Millions of Pounds)

End Products	1975	1977	1980[2]	1985[2]
Acetal resins	55	95	110	140
Acrylic fibers	430	575	620	730
Acrylonitrile-butadiene-styrene (ABS)	635	1020	1250	1730
Butyl rubber	240	310	290	325
Copolyester ethers	5	10	15	35
Epoxy resin (liquid DGEBA)	125	205	250	350
Epoxy resin (solid DGEBA)	60	75	95	140
Ethylene-propylene rubber	35	40	50	65
Ethylene-propylene-diene monomer (EPDM) rubber	155	240	275	360
Melamine-formaldehyde (molding compound)	35	50	65	100
Melamine-formaldehyde (syrup)	110	145	170	230
Modacrylic fiber	35	45	55	70
Nitrile barrier resins (BAREX & CYCOPAC)	90	115	170	325
Nitrile barrier resin (LOPAC)	85	115	170	340
Nitrile rubber	110	135	150	180
Nitrile rubber (latex)	20	25	25	25
Nylon 6 (chips)	40	60	70	115
Nylon 6 (melt)	600	650	735	930
Nylon 66 (chips)	1380	1720	2060	2660
Phenol-formaldehyde (molding compound)	225	325	425	540
Phenol-formaldehyde (syrup)	770	1060	1200	1500
Polyacrylamide	50	80	90	130
Polyacrylate (latex)	95	105	120	150
Polyacrylate (pellets)	235	280	370	500
Polybutadiene	615	800	620	680
Polybutenes	450	550	670	935

Material				
Polybutylene terephthalate (glass-filled)	10	30	55	165
Polybutylene terephthalate (plain)	5	15	20	50
Polycarbonate (general purpose)	60	130	170	235
Polycarbonate (flame-resistant)	25	50	60	80
Polychloroprene (neoprene)	200	260	280	315
Polyether polyol (sorbitol-based hexol)	0	0	0	0
Polyether polyol (phosphorus-containing)	0	0	0	0
Polyether polyol (glycerine-based triol)	0	0	0	0
Polyethylene (high-density)	2085	3160	4070	6175
Polyethylene (low-density)	4410	5860	7070	9960
Polyethylene glycol	95	110	140	225
Polyethylene terephthalate (PET)	3215	3700	4300	5560
Polyethylene terephthalate (PET) barrier resin	0	55	265	520
Polyisobutylenes	120	155	175	205
Polyisoprene	145	135	170	215
Polyisoprene (latex)	5	5	10	15
Polymethyl methacrylate	110	150	170	220
Polypropylene	1640	2435	3200	5600
Polypropylene glycol	95	110	145	230
Polystyrene (crystal grade)	1065	1275	1600	2000
Polystyrene (impact grade)	1135	1780	2100	2840
Polystyrene (expandable beads)	300	360	465	720
Polyurethane flexible foam	1020	1285	1500	1960
Polyurethane rigid foam	340	435	570	1000
Polyvinyl acetate	50	65	80	105
Polyvinyl acetate (latex)	460	510	600	800
Polyvinyl alcohol	55	100	125	195
Polyvinyl butyral	40	40	55	70
Polyvinyl chloride (general purpose)	2920	4220	5400	7700
Polyvinyl chloride (latex)	535	645	860	1400
Styrene-acrylonitrile (SAN)	50	105	120	165

117

Table 4.7 (*Continued*)

End Products	1975	1977	1980[2]	1985[2]
Styrene-block copolymers	55	65	90	140
Styrene-butadiene rubber (SBR)	2505	2765	2750	3000
Styrene-butadiene rubber (latex)	235	255	290	300
Thermoplastic olefin elastomers	15	20	35	65
Thermoplastic polyurethanes	40	50	65	90
Unsaturated polyester (general purpose)	660	890	1000	1260
Unsaturated polyester (corrosion-resistant)	55	80	145	355
Unsaturated polyester (fire-resistant)	45	60	95	190
Urea-formaldehyde (molding compound)	40	60	80	130
Urea-formaldehyde (syrup)	625	905	1100	1400
Vinyl acetate/ethylene copolymer (latex)	40	50	70	100
Vinyl chloride/vinyl acetate copolymer	345	370	415	500
Vinylidene chloride/vinyl chloride copolymer	180	200	230	300
Vinylidene chloride/ethyl acrylate/methyl methacrylate terpolymer (latex)	45	60	90	160

[1] Exports are excluded.
[2] Exogenous demand of end products for 1980 and 1985 are estimated based on the supply/demand patterns for the past several years.

Table 4.8 Updated Actual/Projected U.S. Exogenous Supply/Demand of Intermediate Chemicals and End Products (Millions of Pounds)

Chemicals	1975	1977	1980[1]	1985[1]
Exogenous supply[2,3]				
Acetic acid	35	25	0	0
Polyethylene (high-density)	125	0	0	0
Polyethylene (low-density)	20	20	20	20
Exogenous demand[4]				
Acrylamide	0	0	0	0
Acrylonitrile	0	0	0	0
Adipic acid	75	90	100	135
Bisphenol-A	0	0	0	0
Butadiene	175	245	275	350
1,4-Butanediol	140	175	215	300
Butenes, mixed	0	0	0	0
Caprolactam	0	0	0	0
Chloroprene	0	0	0	0
Cyclohexane	0	0	0	0
Diethylene glycol	310	320	370	440
Dimethyl terephthalate	0	0	0	0
Epichlorohydrin	0	0	0	0
Ethyl acrylate	100	300	400	700
Ethylene	1,180	1,360	1,510	1,980
Ethylene glycol	2,130	1,920	2,000	2,200
Ethylene oxide	1,500	1,700	2,150	2,800
Formaldehyde (100%)	1,100	1,100	1,100	1,100
Glycerine	290	290	290	290
Hexamethylenediamine	0	0	0	0
Isobutylenes	130	150	165	195
Isophthalic acid	90	115	150	230
Isoprene	0	0	0	0
Maleic anhydride	100	130	150	195
Melamine	0	0	0	0
Methanol	1,980	2,625	3,410	4,110
Methyl acrylate	0	0	0	0
Methylene diphenylene diisocyanate	0	0	0	0
Methyl methacrylate	85	155	300	530
Phenol	420	470	500	600
Phosgene	250	285	410	590
Phthalic anhydride	590	775	850	1,040
Propylene, polymer grade	960	1,335	1,510	1,820
Propylene glycol	200	225	240	285
Propylene oxide	90	90	90	120
Sodium chloride	13,870	15,360	17,900	23,000
Styrene	530	550	600	700

119

Table 4.8   *(Continued)*

Chemicals	1975	1977	1980[1]	1985[1]
Terephthalic acid, fiber grade	15	25	30	45
Toluene diisocyanate	0	0	0	0
Urea	7,700	8,500	11,300	18,100
Vinyl acetate	110	145	230	510
Vinyl chloride	105	140	140	140
Vinylidene chloride	0	0	0	0

[1]Exogenous supply/demand of intermediate chemicals and end products for 1980 and 1985 are estimated based on the supply/demand patterns for the past several years.
[2]Imports are excluded.
[3]Exogenous supply of the remainder of end products are zero.
[4]Exports are excluded.

To identify process bottlenecks and to determine capacity expansion needs for the U.S. Industry in 1985, study runs have also been made using the same bound structure as in "Planning the 1985 U.S. Chemical Industry" in Chapter 2 Table 4.11 represents the capacity expansion needs for the polymer processes in 1985, and the utilization factors recommended by the model are seen in Table 4.12.

Acceptance of New Polymer Technologies by the U.S. Chemical Industry in 1985

In this section four potential new polymer processes are introduced to the model. In order to test the attractiveness of these new technologies in the 1985 U.S. chemical industry, the existing capacity utilization factors are again fixed at 50% and 0% of the 1977 built capacities. Table 4.13 shows the new processes added to the model and Table 4.14 represents the major impact of the acceptance of these new technologies on the 1985 industry.

A STUDY OF THERMOPLASTICS INTERCHANGE POTENTIAL

The U.S. economy does not require a specific end product such as polyethylene, polypropylene, or polystyrene; rather, the economy requires the function that such a product can perform, such as being a building block for a variety of industrial and household products (e.g., housewares, tires, clothing). Economic

Table 4.9 Actual/Projected U.S. Process Capacities for End Products (Millions of Pounds)

End Products	Processes[1]	1975	1977	1980[2]	1985[2]
Plastics and resins					
ABS resins	1 + 2 + 3	1265	1445	1900	1900
Barrier resin (nitrile-BAREX)	6	25	30	50	100
Barrier resin (PET)	9	0	55	300	500
	10	0	5	30	60
Epoxy resin (liquid DGEBA)	11 + 12	225	265	350	500
Epoxy resin (solid DGEBA)	13	100	120	150	250
Melamine-formaldehyde (molding compound)	14	55	65	100	150
Melamine-formaldehyde (syrup)	15	170	200	250	300
Nylon 6 (chips)	16	75	90	130	200
Nylon 6 (melt)	17	850	900	900	1000
Nylon 66 (chips)	18	2000	2200	2400	2900
Phenol-formaldehyde (molding compound)	19	400	420	500	600
Phenol-formaldehyde (syrup)	20	1250	1385	1500	1900
Polyacrylamide	21	100	140	150	250
Polybutenes	24	615	750	900	1200
Polyisobutylenes	25	165	210	250	300
PBT (glass-filled, self extinguishing)	26	15	45	100	250
PBT (plain)	27	10	20	30	80
Polycarbonate (general purpose)	28	150	150	150	200
	29	95	95	95	150
Polyether polyol (sorbitol-based hexol)	32	165	220	300	450
Polyether polyol (phosphorus-containing)	33	90	130	150	350
Polyether polyol (glycerine-based triol)	34	980	1165	1600	1800
Polyethylene (high-density)	35	2060	2300	2600	2900
	36	80	460	700	700
	37	180	340	600	600
	38 + 39 + 40	965	1375	1500	2000

Table 4.9 (*Continued*)

End Products	Processes[1]	1975	1977	1980[2]	1985[2]
Polyethylene (low-density)	41	3020	3475	4000	4900
	42	700	800	800	800
	43	2900	3350	4100	4600
Polyethylene terephthalate	45	3390	3730	3800	4100
	46	1630	1900	2300	2400
Polypropylene	49	400	630	900	1000
	50	0	0	200	200
	51	2415	2705	4000	4400
	52	100	140	140	140
	53	0	0	200	200
Polystyrene (crystal grade)	55	1380	1900	2100	(2)
Polystyrene (impact grade)	56 + 57	1800	2655	2700	3500
Polystyrene (expandable beads)	58	465	465	600	800
Polyurethane flexible foam	59	1300	1450	1700	2300
Polyurethane rigid foam	60	600	725	1000	1700
Polyvinyl acetate	61 + 62	70	90	100	150
Polyvinyl alcohol	64	170	210	250	400
Polyvinyl butyral	65	55	60	80	100
PVC (general purpose)	66	325	370	400	400
PVC (latex)	68 + 69	600	680	900	1500
SAN	70 + 71 + 72 + 73	170	290	250	250
Unsaturated polyester (general purpose)	74 + 75	1145	1290	1600	2100
	76 + 77	70	100	180	180
	78	210	350	350	350
Urea-formaldehyde (mold. comp.)	81	60	70	100	150
Urea-formaldehyde (syrup)	82	965	1080	1300	1600
Vinyl chloride/vinyl acetate copolymer	84	480	480	480	600

	Process[1]				
Synthetic fibers					
Acrylic fibers	87	125	125	125	125
	88	70	75	75	75
	89	295	305	310	310
	90	235	295	295	295
Modacrylic fibers	91	10	10	10	10
	92	40	40	40	40
	93 + 94	25	30	30	30
Synthetic rubbers (elastomers)					
Butyl rubber	95	400	450	450	500
EPDM rubber	96	190	235	300	300
	97	100	110	150	—³
EP rubber	98	60	70	70	70
Nitrile rubber	99	185	165	185	185
Nitrile rubber (latex)	100	45	30	30	30
Polybutadiene	101	145	145	145	145
	102	215	300	300	300
	103	145	145	145	145
	104	365	365	365	365
Polychloroprene (neoprene)	105	440	440	440	440
Polyisoprene	107	275	275	—³	—³
Styrene-butadiene rubber (SBR)	110	3100	3080	3000	3000
	111	365	400	400	400
Styrene-butadiene (latex)	112	395	375	375	375
Thermoplastic Elastomers					
Copolyester-ethers	113	10	15	30	60
Thermoplastic polyurethanes	115 + 116	55	55	—³	—³

¹The numbers refer to process numbers in Table 4.6.
²Production capacities for 1980 and 1985 are estimated based on the supply/demand patterns of end products.
³Not constrained.
⁴Production capacity of the remainder of the polymer processes are not constrained.

123

Table 4.10 Process Selection by the Models and the Actual Industry in 1977 (Figures in Millions of Pounds)

End Product	Model I	Model II	Actual
ABS			
1 Via emulsion/emulsion polymerization	—	—	1040
2 Via suspension/emulsion polymerization	1020	—	—
3 Via bulk/suspension polymerization	—	1020	—
4 Net exports	—	—	40
Total[1]	1020	1020	1000
Polyethylene, high-density (HDPE)			
1 Via Phillips technology	2295	2295	2000
2 Others	865	865	1700
3 Net exports	—	—	450
Total[1]	3160	3160	3250
Polyethylene, low-density (LDPE)			
1 Via autoclave reactor	2510	2510	2400
2 Via tubular reactor	3350	3350	4000
3 Net exports	—	—	600
Total[1]	5860	5860	5800
Polyvinyl acetate (PVAc)			
1 Via suspension polymerization	65	65	150
2 Via emulsion polymerization	510	510	430
Total[1]	575	575	580
Polyvinyl chloride (PVC)			
1 Via suspension polymerization	3850	3850	3950
2 Via bulk polymerization	370	370	500
3 Via emulsion polymerization	645	645	750
4 Net exports	—	—	150
Total[1]	4865	4865	5050
SAN			
1 Via bulk polymerization	—	—	—
2 Via continuous emulsion polymerization	—	—	120
3 Via batch emulsion polymerization	—	—	—
4 Via suspension polymerization	105	105	—
5 Net exports	—	—	10
Total[1]	105	105	110

[1]Production − net exports.

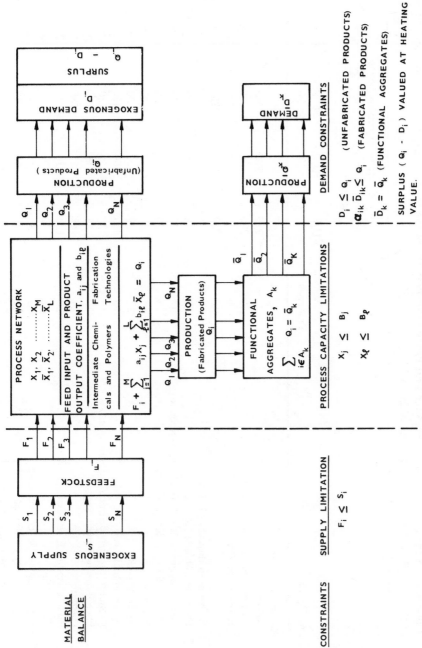

Figure 4.22 Model structure.

125

Table 4.11 Capacity Expansion Needs for End Products in 1985 When the Industry Expands beyond 0 and 50% of the 1977 Built Capacities (Millions of Pounds)

End Products	Processes[1]	Capacity Expansion Needs			
		Model I		Model II	
		U.F. = 50%	U.F. = 0%	U.F. = 50%	U.F. = 0%
Plastics and resins					
ABS resins	1 + 2 + 3	285	285	285	285
Barrier resin (nitrile-BAREX)	6	295	0	0	0
Barrier resin (PET)	9	230	465	230	465
	10	230	0	230	0
Epoxy resin (liquid DGEBA)	11 + 12	85	85	85	85
Epoxy resin (solid DGEBA)	13	20	20	20	20
Melamine-formaldehyde (molding compound)	14	35	35	35	35
Melamine-formaldehyde (syrup)	15	30	30	30	30
Nylon 6 (chips)	16	25	25	25	25
Nylon 6 (melt)	17	30	30	30	30
Nylon 66 (chips)	18	460	460	460	460
Phenol-formaldehyde (molding compound)	19	120	120	120	120
Phenol-formaldehyde (syrup)	20	115	115	115	115
Polybutenes	24	185	185	185	185
PBT (glass-filled, self extinguishing)	26	120	120	120	120
PBT (plain)	27	30	30	30	30
Polycarbonate (general purpose)	28	35	85	35	85
Polyether polyol (sorbitol-based hexol)	32	95	95	95	95
Polyether polyol (glycerine-based triol)	34	300	300	300	300

Material	Process No.[1]				
Polyethylene (high-density)	37	5835	3770	5835	3770
Polyethylene (low-density)	42	9160	5750	9160	5750
Polyethylene terephthalate	46	3660	1795	3660	1795
Polypropylene	50	5600	3900	5600	3900
Polystyrene (crystal grade)	55	150	150	150	150
Polystyrene (impact grade)	56 + 57	185	185	185	185
Polystyrene (expandable beads)	58	255	255	255	255
Polyurethane flexible foam	59	510	510	510	510
Polyurethane rigid foam	60	275	275	275	275
Polyvinyl acetate	61 + 62	15	15	15	15
Polyvinyl butyral	65	10	10	10	10
PVC (general purpose)	66	7330	7330	7330	7330
PVC (latex)	68 + 69	720	720	720	720
Unsaturated polyester (general purpose)	76 + 77	1160	340	1160	340
Urea-formaldehyde (molding compound)	81	60	60	60	60
Urea-formaldehyde (syrup)	82	320	320	320	320
Vinyl chloride/vinyl acetate	84	20	20	20	20
Synthetic fibers					
Acrylic fibers	87	605	270	605	270
Modacrylic fibers	91	60	25	0	0
	92	0	0	30	10
Synthetic rubbers					
EPDM rubber	97	305	185	305	185
Nitrile rubber	99	15	15	15	15
Polybutadiene	102	645	320	525	200
SBR	111	2600	1060	0	0
Thermoplastic elastomers					
Copolyester-ethers	113	20	20	20	20
Thermoplastic polyurethanes	115 + 116	35	35	35	35

[1]The numbers refer to process numbers in table 4.6.

Table 4.12 Recommended Utilization Factors of the Polymer Processes in 1985 When the Industry Expands beyond 0 and 50% of the 1977 Built Capacities

End Products	Processes[1]	Model I U.F. = 50%	Model I U.F. = 0%	Model II U.F. = 50%	Model II U.F. = 0%
Plastics and resins					
Barrier resin (nitrile-BAREX)	6	—[2]	0	50[3]	0
Barrier resin (PET)	10	—[2]	0	—[2]	0
Polyacrylamide	21	93	93	93	93
Polyisobutylenes	25	98	98	98	98
Polyether polyol (phosphorus-containing)	33	94	94	94	94
Polyvinyl alcohol	64	94	94	94	94
SAN	70 + 71 + 72 + 73	57	57	57	57
Synthetic fibers					
Modacrylic fibers	91	50[3]	0	—[2]	—[2]
	92	—[2]	—[2]	50[3]	0
Synthetic rubbers					
Butyl rubber	95	72	72	72	72
EP rubber	98	93	93	93	93
Nitrile rubber (latex)	100	83	83	83	83
Polychloroprene	105	50[3]	72	50[3]	0
SBR	110	91	97	50[3]	0
	111	50[3]	0	—[2]	—[2]
SB (latex)	112	80	98	98	98

[1] The numbers refer to process numbers in Table 4.6.
[2] Expansion needed.
[3] Forced to the industry.

Table 4.13 Potential New Polymer Processes Accepted in 1985 When the Industry Expands beyond 0 and 50% of the 1977 Built Capacities

Process Number	Process Description	Model I U.F. = 50%	Model I U.F. = 0%	Model II U.F. = 50%	Model II U.F. = 0%
NP1	Polyethylene (high-density) by a gas phase process	—	—	—	—
NP2	Polyethylene (high-density) via Mitsubishi technology	—	—	—	—
NP3	Polyethylene (low-density) by a gas phase process	X	X	X	X
NP4	Polyvinyl chloride (general purpose) by vapor phase bulk process	X	X	X	X

Table 4.14 Impact of New Polymer Processes on the Capacity Expansion Needs and Utilization Factors of the Processes in 1985 When the Industry Expands beyond 0 and 50% of the 1977 Built Capacities

End Products	Processes[1]	Capacity Expansion Needs (Millions of Pounds)				Utilization Factors			
		Model I		Model II		Model I		Model II	
		U.F. = 50%	U.F. = 0%	U.F. = 50%	U.F. = 0%	U.F. = 50%	U.F. = 0%	U.F. = 50%	U.F. = 0%
Polyethylene (low-density)	42	0	0	0	0	50[2]	0	50[2]	0
	NP3	6150	9960	6150	9960	—[3]	—[3]	—[3]	—[3]
PVC (general purpose)	66	0	0	0	0	50[2]	0	50[2]	0
	NP4	7515	7700	7515	7700	—[3]	—[3]	—[3]	—[3]

[1]The numbers refer to process numbers in Tables 4.6 and 4.14.
[2]Forced to the industry.
[3]Expansion needed.

studies can forecast changes in the demand for such use areas, but the forecast is more difficult for the particular cast of chemical building blocks needed to serve these functions in the economy. Therefore, capability should exist to determine the products that the chemical industry ought to manufacture and the technology that should be used to meet future needs of the economy. As a result, the model is extended to include functional aggregates in which substitutions can be made among chemical products, if necessary, to achieve a minimum total production cost to the industry. Some additional terms are therefore added to the model and shown in Figure 4.22.

The production cost to the industry is defined as

$$\text{production cost} = \sum_{i=1}^{N} P_i F_i + \sum_{j=1}^{M} C_j X_j + \sum_{l=1}^{L} C_l \overline{X}_l + \sum_{i=1}^{N} (Q_i - D_i)(P_i - H_i)$$

In general, chemical products are fabricated at levels $\overline{X}_l$ (with the specified capacity limits B_l) with the total production cost of C_l. These finished products are responsible for satisfying the demand for functional aggregates $\overline{D}_k$, which are needed by the consumer market. The factor α_{ik} (in the range of 0 to 1) is introduced to the model, representing the fraction of finished product i that is required in the functional aggregate k. The factor α_{ik} in fact controls the presence of a specific end product with the desired quality in a certain use area; $\alpha_{ik} = 0\%$ represents a situation in which all of the finished product i can be substituted by the competing products in functional aggregate. As α_{ik} increases, the industry would have less freedom to select the most economical finished product to meet the needs of the economy, and at $\alpha_{ik} = 100\%$ all of the actual/projected demand for the fabricated product i in functional aggregate k, $\overline{D}_{ik}$, has to be satisfied by the industry.

To show potential uses of the model, allowable substitutions that can be made among chemical products in pipe and tubing functional aggregate are studied. As shown in Figure 4.8, high- and low-density polyethylene, impact grade polystyrene, polypropylene, polyvinyl chloride, and ABS (acrylonitrile-butadiene-styrene) are the major ingredients of this functional aggregate. Parametric studies are made based on the 1977 and 1985 U.S. chemical industry; the industry is assumed to have the freedom to reach its optimal structure within the specified capacity limitations and all the substitutions are to be made based on *resin* availability and price. In other words, process capacities are specified as upper limits and the fabrication technologies have not been considered. Figures 4.23, 4.24, 4.25, and 4.26 represent the run results for both years when the factor α_{ik} is varied between 0% and 100%. It is evident that high-density polyethylene is the most attractive feedstock for the manufacture of pipe and tubing. However, capacity limitations on this chemical forces the model to select the other end products such as low-density polyethylene and polyvinyl chloride in order to satisfy the consumer's demand for plastic pipes.

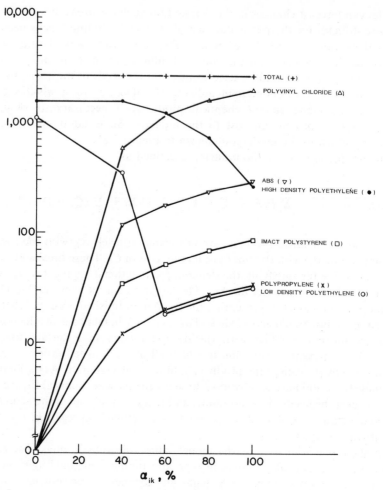

Figure 4.23 Thermoplastics interchange potential within pipe and tubing functional aggregate in 1977—Model I (millions of pounds).

By relaxing the capacity limitations and allowing the industry to have the total freedom in selecting the most economical resin to manufacture this functional aggregate ($\alpha_{ik} = 0\%$), high-density polyethylene becomes responsible for satisfying the total demand for pipe and tubing.

The major impact of selection of polyethylene for this functional aggregate on technology selection of the industry is focused on the ethylene manufacturing processes. Extensive production of ethylene for polyethylene results in great amounts of propylene as a side-product of the steam cracking processes.

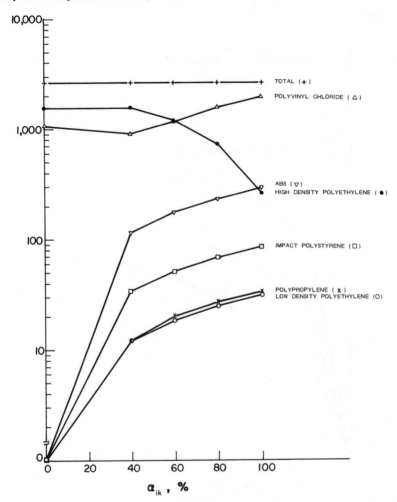

Figure 4.24   Thermoplastics interchange potential within pipe and tubing functional aggregate in 1977—Model II (millions of pounds).

However, industry prefers not to consume this chemical to manufacture polypropylene-based pipe and tubing, but rather a switch occurs in steam cracking processes. In other words, to prevent having surplus propylene, such heavy feedstocks as gas oil and/or naptha lose their attractiveness for ethylene production and instead ethane and/or propane become relatively more attractive feedstocks for steam cracking processes because of lower propylene yields.

It is evident from the results that fabrication technologies for pipe and tubing are in fact the deciding factors for product allocation in this functional ag-

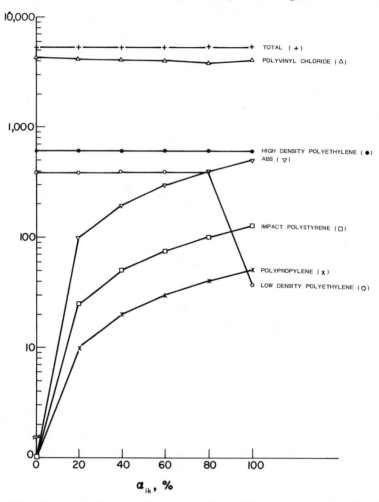

Figure 4.25 Thermoplastics interchange potential within pipe and tubing functional aggregate in 1985—Model I (millions of pounds).

gregate. However, based on resin price only, high-density polyethylene is the best candidate for production of plastic pipes.

Product quality is another deciding factor in resource allocation and technology selection in the chemical industry. For example, production of polyethylene terephthalate (PET) resins is actually controlled by the quality of the fabricated product. Over 90% of PET resins which are made from either dimethyl terephthalate (DMT) or terephthalic acid (TPA) and ethylene glycol are used in fiber manufacturing in clothing, home furnishings, tire cord, and so on. TPA-based resins have several advantages over DMT-based resins such as

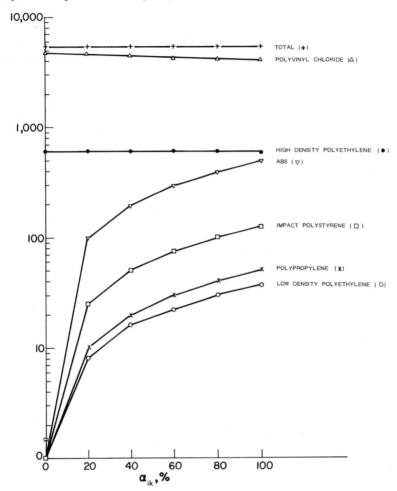

Figure 4.26 Thermoplastics interchange potential within pipe and tubing functional aggregate in 1985—Model II (millions of pounds).

faster reaction, low catalyst residue in the polymer, a by-product (water) that does not require special recovery facilities, and so on. But, they have dye uniformity problems in the fiber manufacturing step. Therefore more than 70% of all the polyester fibers produced in the United States are based on DMT. Clearly, TPA would have been the most attractive feedstock for the manufacture of polyester fibers, if the resin price was the only deciding factor. This is also evident from the model run results. In other words, when the industry is allowed to reach its optimal structure with no capacity limitations, TPA takes all the load off DMT for the manufacture of PET resins.

However, for more conclusive studies of substitutions that can be made among chemical products, inclusion of fabrication technologies seems to be necessary and the model constructed here is the first step toward this goal.

CONCLUDING REMARKS

The demand for specific properties in end products drives technical changes within the intermediate chemicals industry. For many changes in polymeric products, little impact is felt in the intermediate industry, because a wide variety of polymers derive from the same monomers. However, several changes within functional aggregates have major impacts and for this reason a comprehensive study of chemical products is essential.

CHAPTER FIVE

Pricing Structure
of the U.S. Chemical Industry

The forces of the market place tend to adjust the prices of chemicals to the point where prices accurately measure value in the economy. If the Economic Model I and the Integrated Model II are the accurate perceptions of an optimal industry, an important event should occur. The *shadow prices* of Model II would be directly related to the market prices of Model I. In this chapter, we examine the implications of this and show how the introduction of new technologies can change the industrial system sufficiently to confound price projections and confuse decision making. We also show how the pricing structure of the industry changes in response to the increased restrictions on the use of toxic chemicals.

CHEMICAL PRICES

The Integrated Model is a primal linear program whose elements correspond to the stoichiometric, raw material and finished molecule constraints, the processing costs and the capacity allocation. The dual of that establishes shadow prices (dual activities) that define the local economic criteria to be used to identify sites for process modification, in the same way that purchase and sales prices are used, along with investment costs, to establish the economic attractiveness of proposed engineering activities. Should the primal linear program be defined as:

$$\text{Min } cx$$
$$Ax \geq d$$
$$x \geq 0$$

the corresponding dual is

$$\text{Max } yd$$
$$yA \leq c$$
$$y \geq 0$$

From the duality theorem, if either the primal or the dual problem has a finite optimum solution, then the other problem has a finite optimum solution and the extremes of the linear functions are equal. That is, $cx^\circ = y^\circ d$, where x° and y° are the solution vectors to the primal and the dual problem, respectively.

However, it can be visualized that the chemicals can either be manufactured by the petrochemical industry at a minimum cost of cx°, or be obtained from an hypothetical market with the system of maximum prices of y° at the same cost to the economy. Therefore, one can conclude that for a well-integrated and well-balanced economy, shadow prices would indeed be a good representation of the chemicals' market values.

Using the 1977 economic environment, part of the system of prices obtained from the model industry is summarized in Table 5.1. It is clear that the market prices, in general, have reasonably good agreement with the shadow prices obtained from the dual problem. However, for some chemicals, such as hexamethylenediamine (HMDA), hydrogen cyanide, and propylene oxide, substantial price changes are shown; this is justified as follows:

1   Because of hydrogen cyanide capacity limitations, the most attractive process for hexamethylenediamine, the butadiene route, cannot be operated at full capacity. The alternate processes such as acrylonitrile and adipic acid-based processes are also used to produce HMDA. By allowing the model industry to operate ideally with no process capacity limitations, the total consumption of hydrogen cyanide increases by 140% and the shadow prices for hydrogen cyanide and HMDA decrease by 61% and 33%, respectively.

2   Hydroperoxide process to styrene (Halcon's technology) resulting in propylene oxide as a co-product is the most attractive process for the manufacture of propylene oxide. But due to the capacity constraints, the model industry has to utilize such processes as chlorohydration and/or oxidation of propylene to satisfy the needs for propylene oxide. However, by operating the industry with no capacity constraints, the hydroperoxide process becomes the only source for the manufacture of propylene oxide and therefore a drop of 49% in its shadow price is observed.

Because chemical prices are the direct reflections of perturbations in the supply/demand pattern and available manufacturing capacity, acceptance of the new technologies would change such a pattern and therefore create new prices.

Fourteen potential new technologies of Tables 2.11 and 4.13 were again introduced to the Integrated Model and their acceptance was tested by the 1985 industry. Using the 1985 economic environment, Table 5.2 represents the processes accepted by the model industry and Table 5.3 shows the major changes in shadow prices resulting from the acceptance of such technologies. Table 5.4 also represents the cost improvement (reduced cost) required for the new technologies to be accepted by the projected industry. It is recognized that the reduced costs (RC) have the following relation with the shadow prices:

$$(RC)_j = C_j{}^* - \sum_{i=1}^{N_j} a_{ij} \, |P_i{}^*| \qquad (1)$$

where N_j is the total number of chemicals participating in process j (including the main product m), and P_i^* is the shadow price of chemical i. By rearranging equation (1), the following can be obtained:

$$|P_m^*| = C_j - (RC)_j \qquad (2)$$

where $C_j = C_j^* - \sum_{\substack{i=1 \\ i \neq m}}^{N_j-1} a_{ij}|P_i^*| = $ total manufacturing cost of process j

The processes which appear in the basis of the linear program, meaning the most efficient technologies, have the reduced cost of zero. Therefore, from

Table 5.1 The System of Prices Obtained From the Model Industry in 1977 (¢/lb)

| | | Shadow Prices[1] | |
Chemicals	Actual Prices	Model Industry	Ideal Industry[2]
Acetaldehyde	16.5	13.82	13.46
Acetic acid	16	13.63	11.63
Acetic anhydride	25	22.36	19.84
Adipic acid	28	32.41	26.37
Ammonia	7.5	8.64	8.64
Aniline	27	27.5	27.41
Bisphenol-A	40	41.04	40.39
Butadiene	19.3	22.86	22.86
Cyclohexane	14	13.55	13.45
Cyclohexanone	31	29.53	26.57
Dimethyl Terephthalate	31	31.54	28.58
Ethanol	17	17.51	18.36
Ethylbenzene	13	11.68	11.99
Formaldehyde	14	15.38	15.38
Hexamethylenediamine	48	80.53	53.82
Hydrogen cyanide	25	71.55	27.91
Methanol	7	7.37	7.37
Methyl Ethyl Ketone	18.5	16.11	20.30
Nitrobenzene	14.6	15.19	15.19
Phenol	23	19.64	21.25
Phthalic anhydride	24	24.18	24.18
Propylene oxide	22.5	54.64	27.74
Styrene	19.5	18.12	18.47
Urea	7.8	9.66	8.81
Vinyl acetate	23	23.32	22.36

[1]Absolute values.
[2]The same as the Model Industry except that the processing capacities were removed.

Table 5.2   New Technologies Accepted in 1985 When the Industry
Expands beyond 0 and 50% of the 1977 Built Capacities

Chemical	Process[1]	Utilization Factor	
		0%	50%
Aniline	N1		
	N2	X	X
Ethylene glycol	N3		
	N4		
Maleic anhydride	N5		X
Phenol	N6	X	X
Styrene	N7		
	N8		
	N9	X	X
Vinyl acetate	N10	X	X
Polyethylene (H.D.)	NP1		
	NP2		
Polyethylene (L.D.)	NP3	X	X
Polyvinyl chloride	NP4	X	X

[1]The numbers refer to the process numbers in Tables 2.11 and 4.13.

equation (2), the total manufacturing cost of such technologies would indeed be equal to the shadow price of the corresponding main product. For a positive reduced cost, the resulting product would have to be charged at a rate lower than the manufacturing cost to compete for the consumer market, and finally a process would show a negative reduced cost if there were incentives for capacity expansion, resulting in a situation in which the main product became too valuable and it would show a shadow price higher than the total manufacturing cost.

However, it is observed that the new styrene process based on toluene is very attractive in both U.F. = 0% and 50%. But when the model industry is forced to operate at 50% U.F., substantial producing capacity of the ethylbenzene dehydrogenation process to styrene is forced into the model, reducing the operating level of the Halcon's more modern process and therefore losing substantial quantities of propylene oxide co-product. As a result, such processes as chlorination and/or oxidation of propylene to propylene oxide have to be operated at higher levels to make up for the needs for this chemical and this results in a relatively more expensive propylene oxide (higher shadow price).

Similar kinds of observations were made in the production pattern of phenol, acetone, cumene, and hydrogen peroxide. When the model industry operates at 0% U.F., the new phenol process based on toluene becomes very attractive, which in turn, tends to replace the cumene-based process. However, due to the needs for acetone, the isopropanol-based hydrogen peroxide becomes the best producer of acetone as by-product. But because of the limited acetone production from this process, the cumene route to phenol is also used as an acetone supplier. Because the resulting phenol from this process has to compete with the toluene-based phenol in the same market, acetone has to be charged at a higher price which in turn results in cheaper hydrogen peroxide. By operating the industry at 50% U.F., substantial quantities of acetone become available from the cumene-based phenol process which results in a price drop for this chemical. However, the resulting phenol from this process would still have to compete with the toluene-based phenol, therefore, the price of cumene has to be lowered artificially by about 20%. In other words, although the total manufacturing cost of cumene remains unchanged, it has to be available at a substantially lower price to enter the market and be accepted by the phenol manufacturers.

IMPACT OF RESTRICTIONS ON TOXIC SUBSTANCES ON THE PRODUCTION OF SYNTHETIC MATERIALS

The Toxic Substances Control Act (TSCA) has started to have significant impact on the chemical industry. This act gives the Environmental Protection Agency (EPA) far-reaching authority to regulate the manufacture, processing, use, and disposal of chemicals perceived to be dangerous to the health of mankind and/or to the environment. According to the current list of chemical substances for testing for health and environmental effects (Table 5.5), benzene-based chemicals are exposed to intense scrutiny and decisions made regarding such chemicals would have serious effects on growth and development of certain sectors of the chemical industry.

Restrictions on the use of the toxic chemicals may indeed influence the choice of process routes to commercial products or the development of new uses for existing products and/or new products to displace more hazardous ones. Because the chemical prices are a combination of the consumer demand, availability, and the function in the economy, such restrictions would sufficiently disrupt the supply and demand balances in the economy and therefore create new pricing structures.

The Role of Benzene in the U.S. Petrochemical Industry

In this section the role of benzene in the U.S. petrochemical industry is examined, using the 1985 economic environment and assuming that the industry has the total freedom, regardless of the installed capacities, to select the optimal technologies in order to minimize its total production cost (Case 1). Parametric studies are made on the price of benzene. To this effect, a factor α is defined, representing percentage increase in the price of benzene produced and/or consumed by the model industry. Changing this factor represents different economic environments and allows one to observe the industry's behavior

Table 5.3 Impact of Acceptance of the New Technologies on the System of Prices in 1985 When the Industry Expands beyond 0 and 50% of the 1977 Built Capacities

Chemicals	Shadow Prices[1] (¢/lb)			
	No New Technology Added		14 New Technologies Added	
	U.F. = 0%	U.F. = 50%	U.F. = 0%	U.F. = 50%
Acetic acid	21.18	21.18	23.81	21.18
Acetone	36.86	37.46	46.41	37.46
Aniline	49.40	49.40	43.48	41.16
Benzene	19.22	19.22	19.00	19.00
Bisphenol-A	71.68	71.85	69.03	64.68
Cumene	30.11	30.38	30.00	24.42
Ethylbenzene	22.10	21.97	21.94	21.81
Ethylene	20.24	19.73	20.22	19.73
Hydrogen peroxide	76.18	81.28	56.20	81.28
Isopropanol	34.88	37.70	34.96	37.70
Maleic anhydride	60.97	60.97	60.82	60.62
Nitrobenzene	27.19	27.19	27.05	27.05
Phenol	40.66	40.66	34.69	32.61
Polyisocyanates (MDI/PMPPI)	99.15	99.15	95.49	92.54
Propylene (chemical grade)	32.48	36.34	32.58	36.34
Propylene oxide	47.72	50.82	58.61	67.42
Styrene	34.08	33.92	29.46	26.94
Toluene	12.40	12.40	14.19	12.40
Vinyl Acetate	41.41	41.21	37.95	34.83
Vinyl chloride	22.44	22.17	22.43	22.17

ABS resins	78.32	77.99	75.22	73.30
Epoxy resin (liquid DGEBA)	119.22	119.32	117.26	114.02
Epoxy resin (solid DGEBA)	119.42	119.49	117.44	114.05
Phenol-formaldehyde (molding compound)	105.99	105.99	100.61	98.75
Phenol-formaldehyde (syrup)	57.70	57.70	52.75	51.03
Polycarbonate (general purpose)	174.56	174.71	172.69	168.19
Polycarbonate (flame-resistant)	180.34	180.48	178.69	174.60
Polyethylene (L.D.)	54.83	54.31	38.66	38.18
Polystyrene (crystal grade)	48.67	48.51	43.96	41.39
Polystyrene (expandable beads)	76.40	76.24	71.86	69.39
Polystyrene (impact grade)	60.32	60.06	55.79	53.21
Polyvinyl acetate	63.92	63.71	60.39	57.19
Polyvinyl acetate (latex)	63.62	63.42	60.09	56.90
Polyvinyl alcohol	135.07	134.70	124.78	122.06
Polyvinyl butyral	204.25	205.29	199.90	196.34
Polyvinyl chloride (general purpose)	46.80	46.52	37.50	37.23
SAN	63.78	63.56	60.36	58.39
Vinyl acetate/ethylene copolymer	68.30	68.03	64.77	61.84

[1]Absolute values.

Table 5.4 Cost Improvement Required for New Technologies to be Accepted in 1985 When the Industry Expands beyond 0 and 50% of the 1977 Built Capacities

Chemical	Process[1]	Cost Improvement (¢/lb)	
		U.F. = 0%	U.F. = 50%
Aniline	N1	5.16	7.67
	N2	0	0
Ethylene glycol	N3	30.21	29.16
	N4	28.74	28.51
Maleic anhydride	N5	2.52	0
Phenol	N6	0	0
Styrene	N7	40.18	40.07
	N8	2.97	3.69
	N9	0	0
Vinyl acetate	N10	0	0
Polyethylene (H.D.)	NP1	1.30	1.30
	NP2	1.15	1.18
Polyethylene (L.D.)	NP3	0	0
Polyvinyl chloride (general purpose)	NP4	0	0

[1]The numbers refer to the process numbers in Tables 2.11 and 4.13.

with respect to increased restrictions on the production of benzene. Figure 5.1 shows the industry's response to production of this chemical feedstock at different α values, and Figure 5.2 represents the increase in the total production cost of the industry. It is observed that, in general, an exogenous supply of chemicals that could also be manufactured from benzene is preferred over the internal production, at different α values. For example, at higher prices of benzene, the model industry prefers to consume an exogenous supply of such intermediate chemicals as adipic acid, cyclohexane, cyclohexanol, cyclohexanone, ethylbenzene, and phenol, and this clearly results in reductions in the total consumption of benzene. However, because such chemicals are not supplied to the model industry in great quantities, the structure of the industry remains unchanged and at reasonable values of α no changes occur in process routes that lead to commercial products. This is also reflected in the shadow prices of the chemicals that involve the consumption of benzene. Figures 5.3 and 5.4 show the impact of the price increase for benzene on the pricing structure of the industry. Because no significant disruptions occur in the supply and demand balances, the shadow prices vary linearly as the price of benzene increases. For the values of α over 130%, the price of benzene becomes high enough to affect the performance of the polymer products industry. For example, at $\alpha = 135\%$ a shift occurs in processes for the manufacture of nitrile bar-

Table 5.5 Priority Testing Recommendations Under TSCA

First List of Chemicals and Categories

Alkyl epoxides
Alkyl phthalates
Chlorinated benzenes, mono- and di-
  chlorinated paraffins (35–70% Cl)
Chloromethane
Cresols
Hexachloro-1,3-butadiene
Nitrobenzene
Toluene
Xylenes

Second List of Chemicals and Categories

Acrylamide
Aryl phosphates
Chlorinated naphthalenes
Dichloromethane
Halogenated alkyl epoxides
Polychlorinated terphenyls
Pyridine
1,1,1-Trichloroethane

Third List of Chemicals and Categories

Chlorinated benzenes, Tri-, Tetra-, Penta-1,2-dichloropropane
Glycidol and its derivatives

rier resins Barex and Cycopac which are used in the food packaging industry. It is observed that at this point Barex becomes the preferred product to avoid the consumption of styrene which is used in Cycopac-type resin. Therefore, further reduction in consumption of benzene is shown. Technological changes of this kind also occur among the high-volume polymer products such as ABS and SBR. For instance, the results show that at an α value over 300%, the suspension/emulsion polymerization process to ABS replaces the previously preferred bulk/suspension polymerization process, simply due to the lower consumption of styrene. Similarly, at an α value over 400%, SBR via emulsion polymerization becomes superior over the solution polymerization route.

However, it is realized that benzene is such a valuable and established chemical that the industry shows a great resistance to reducing its use even at an α value as high as 1000% which causes only about 15% reduction in the total consumption of benzene. This point corresponds to an asymptotic value of about 12 billion pounds of benzene consumed by the chemicals industry in 1985. Therefore, one can conclude that with the present structure of the in-

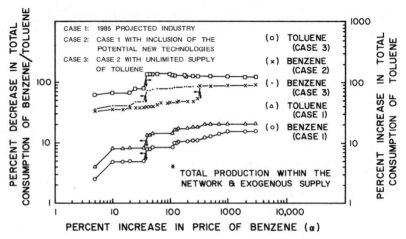

Figure 5.1 Impact of price increase for benzene on the total consumption* of benzene and toluene in 1985.

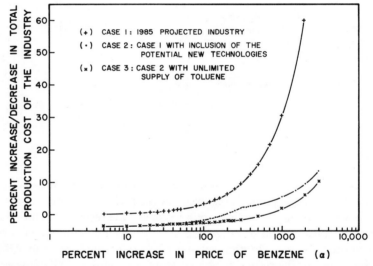

Figure 5.2 Impact of price increase for benzene on the total production cost of the industry in 1985.

dustry not much can be done to avoid the use of benzene. Therefore, to lower the consumption of this chemical, the petrochemical industry has to either employ alternate feedstocks which are less toxic or substitutions must occur among polymer products.

Recently toluene has received considerable attention as a replacement for benzene, because it is produced in much larger quantities. Also toxicologically, toluene may be a less dangerous alternative. However, acceptance of toluene by

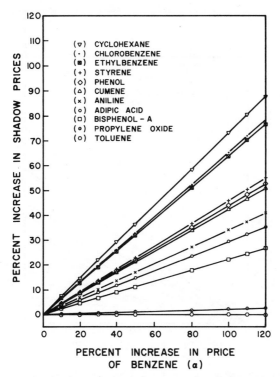

Figure 5.3 Impact of price increase for benzene on the shadow prices of intermediate chemicals in 1985—Case 1.

the industry can significantly disrupt the supply and demand balances for benzene and therefore create new pricing structures for some sectors of the industry. To illustrate this, the potential new technologies outlined in Tables 2.11 and 4.13 were again introduced to the 1985 industry (Case 2), because there are four new toluene-based technologies for aniline, phenol, and styrene.

However, because of the acceptance of such technologies by the projected industry (Table 5.2), the total consumption of benzene drops by more than 27% and the total consumption of toluene becomes limited to its maximum exogenous supply and this represents an increase of more than 44%.

Parametric studies were again conducted on the price of benzene. The impact of these sensitivity studies on the total consumption of benzene and the total production cost of the industry is represented in Figures 5.1 and 5.2. It is observed that the industry shows less dependency on benzene than before simply because toluene acts as an alternate feedstock in this case. This is reflected in the selection of technologies and their production level. For example, for the values of α between 5 and 25%, the new toluene-based phenol process becomes even more attractive and forces the cumene-based route to operate at only 19% of its 1977 installed capacity. This, in turn, forces the cumene

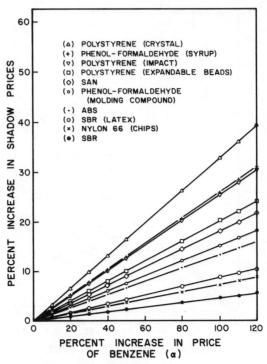

Figure 5.4 Impact of price increase for benzene on the shadow prices of polymer products in 1985—Case 1.

manufacturers to reduce their operating level to little more than 18% of the 1977 built capacities. As a result of these transformations, the chemical industry would have to operate the propylene oxidation route to acetone at higher production levels to meet the demand for this chemical. When the price of benzene is increased further, the toluene-based phenol becomes even more attractive and at $\alpha = 30\%$ this process replaces the cumene-based route and limits the cumene production to only 3% of the 1977 built capacity, which is just enough to meet the exogenous demand of this chemical. As a result, further expansion is required for the propylene oxidation route to acetone. Because all of these changes require a relatively higher amount of toluene and the availability of this chemical feedstock was already limited to its maximum exogenous supply, the production of styrene via toluene dimerization and disproportionation of stilbene (process N9) drops by about 21%. Therefore, to meet the demand for styrene, the Halcon's hydroperoxide process has to be further expanded which results in extensive production of propylene oxide as a side-product which, in turn, reduces the price of this chemical. The processing structure of the industry remains relatively unchanged for higher values of α up to 60%. However, as the price of benzene increases, there are more incentives for toluene to be used in larger quantities. This is also shown in Figures 5.5–5.7

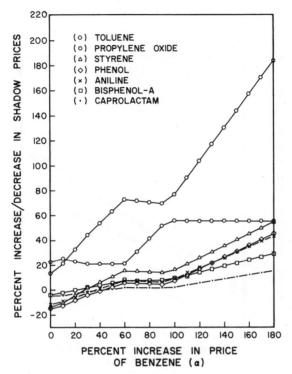

Figure 5.5   Impact of price increase for benzene on the shadow prices of intermediate chemicals in 1985—Case 2.

which represent the changes in pricing structure of the chemicals which involve the consumption of benzene or toluene. As seen, due to the disruptions in supply and demand balances for toluene, the price of this feedstock shows a steady increase for the lower values of α. Increasing the price of benzene makes toluene an even more desirable feedstock and at an α value of 70%, toluene becomes so valuable that the other toluene-based styrene process (process N8) becomes economically acceptable by the model industry and it even shows a higher production level than process N9. This is simply because this new process (N8) offers a better yield than process N9. This transformation results in a relatively more available toluene and therefore a slight price decrease for this feedstock is observed (Figure 5.5). Because the structure of the industry remains the same for 70% $\leq \alpha \leq$ 90%, the price of toluene remains relatively unchanged. However, in this range of α, the resulting styrene from the Halcon's hydroperoxide process has to compete with the toluene-based styrene in the same market. Because the price of ethylbenzene (the feedstock for the Halcon's technology) shows a steady increase due to the higher prices (Figure 5.6); therefore, propylene oxide side-product from the Halcon's process has to be credited at higher values to generate a competitive price for the resulting

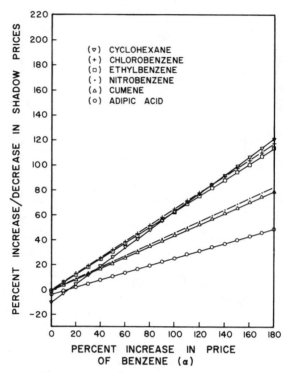

Figure 5.6 Impact of price increase for benzene on the shadow prices of intermediate chemicals in 1985—Case 2.

styrene. At $\alpha = 100\%$, the new styrene process based on toluene and carbon monoxide (process N8) completely replaces the other toluene-based styrene (process N9) and therefore toluene continues to be more expensive as the price of benzene increases. This, therefore, results in a relatively stable pricing structure for propylene oxide. However, no significant changes occur in the processing structure of the industry as α is increased to 200%. But at $\alpha = 210\%$, similar kinds of transformations as Case 1 occur in the polymer products industry, and at an α value over 250% the extremely high prices of benzene and toluene cause a major breakpoint in the industry. At such an economic environment the new butadiene-based styrene (process N7) becomes competitive, and at an α value over 300% this process forces the Halcon's technology to operate at only 46% of its 1977 installed capacity. This transformation drastically reduces the total consumption of benzene (Figure 5.1) and causes a breakpoint in the total production cost of the industry (Figure 5.2).

It is, however, realized that by acceptance of the potential new technologies outlined in Tables 2.11 and 4.13, the industry shows much less dependency on benzene and it is observed that at extremely high values of α the total consumption of benzene can be reduced to its minimum level of about 1.7 billion pounds.

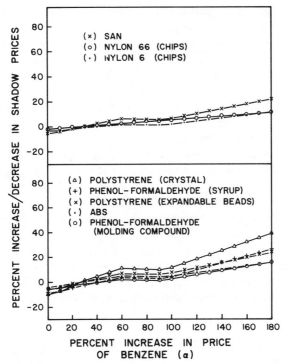

Figure 5.7   Impact of price increase for benzene on the shadow prices of polymer products in 1985—Case 2.

In Case 2, toluene acted as a potential substitute for benzene. However, the chemical industry had to adapt itself to the limited supply of this chemical feedstock.

Coal carbonization processes were once the primary source of toluene. Growing scarcity of crude oil and natural gas has directed attention to coal as a supplier of carbon atoms necessary for petrochemicals. This, therefore, may result in higher supplies for toluene. To this effect, another case is constructed (Case 3) which uses the same bases as Case 2 except that it is also asssumed that an unlimited supply of toluene will be available to the 1985 industry. Although this is a rather extreme case, it serves the purpose of this study. The major impact of having an unlimited supply of toluene is focused on the production/consumption pattern of phenol, cumene, and acetone. It is observed that (at $\alpha = 0\%$, Case 3), the new toluene-based phenol becomes even more attractive and it completely replaces the cumene-based technology and therefore limits the cumene production to its exogenous demand. As a result, further expansion is required for the propylene oxidation route to acetone. It should be noted that this corresponds to the similar kinds of observations made in Case 2 at an α value of 30%. As a result of these changes, the total consumption of benzene drops by more than 32% and the total consumption of toluene increases by about 63% (using Case 1 as the basis for comparison).

The performance of the industry under different pricing structures for benzene is again studied. Figures 5.1 and 5.2 show the impact of price increases for benzene on the total consumption of this chemical and toluene, and the total production cost of the industry, respectively. The major changes in the pricing structures of the industry are also represented in Figures 5.8 and 5.9. It should be noted that in all of these figures, the 1985 projected industry (at $\alpha = 0\%$) is used as the basis for comparison.

In this case, because an unlimited amount of toluene is assumed to be accessible to the 1985 industry, acceptance of the new technologies would not result in disruptions in the supply and demand balance for this chemical feedstock and therefore no change in its shadow price is observed. This is also reflected on the pricing structures for aniline and phenol at different values of α, which in turn, results in relatively unchanged prices for bisphenol-A and caprolactam. For the values of $\alpha = 0$–20%, no significant changes occur in the structure of the industry. However, when the α is increased to 25%, Halcon's process to styrene starts to lose its attractiveness and in fact the production of styrene from this process drops by more than 15%. For the α values up to 35%

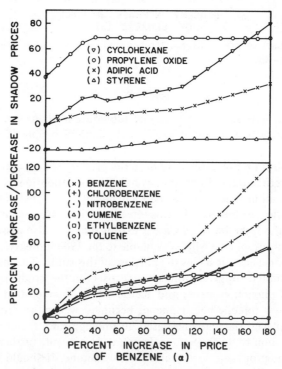

Figure 5.8 Impact of price increase for benzene on the shadow prices of intermediate chemicals in 1985—Case 3.

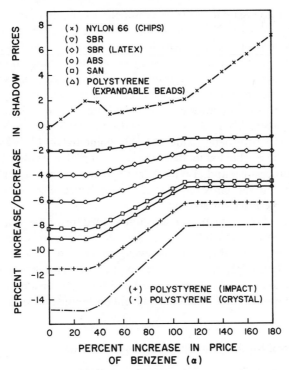

Figure 5.9 Impact of price increase for benzene on the shadow prices of polymer products in 1985—Case 3.

the price of styrene remains relatively unchanged, because substantial quantities of styrene are produced by the new toluene-based styrene process (process N9) which creates a stable price for this chemical. This, however, results in higher prices for propylene oxide, which is produced mainly as the byproduct of the Halcon's hydroperoxide process, to offset the increased raw material cost (mainly ethylbenzene) of this process. At $\alpha = 40\%$, the Halcon's process becomes responsible for only 16% of the total production of styrene and the total consumption of benzene drops by about 74% (Figure 5.1). It is realized that at this relatively low value of α, the industry has significantly departed from benzene and the producers of this chemical can no longer compete with the alternate toluene feedstock, if the price of benzene grows at the same rate. Therefore at higher α values the benzene suppliers have to artificially lower the price in order to compete with the toluene-based technologies. Because substantial quantities of benzene are obtained as the by-product of the new toluene-based styrene process N9, the price adjustment for benzene results in a relatively more expensive styrene, but the price of propylene oxide remains unchanged due to the higher prices for ethylbenzene. However, when α reaches 120%, the adjusted price of benzene still becomes high enough to force the benzene alkylation process to ethylbenzene out of the optimal industry but not sufficiently high to create a stable environment for the manufacture of styrene. In other words, at $\alpha = 120\%$, the other new styrene process based on toluene and carbon monoxide (process N8) becomes attractive and in fact it shares about 24% of all the styrene produced by the industry. Small quantities of by-product ethylbenzene from this process plus the exogenous supply of this chemical would be enough to meet the exogenous demand of ethylbenzene and it therefore reduces the operating level of the Halcon's technology to 45% of its 1977 built capacity. This, therefore, results in stable pricing structures for ethylbenzene and propylene oxide. At this point ($\alpha = 120\%$), the price of benzene has to be readjusted to offset the raw material cost (higher toluene consumption) of process N9 and also to offset the increased investment costs due to the restrictions on the emission standards for benzene in order to create a unique and stable price for styrene.

The structure of the industry remains unchanged for $120\% \leq \alpha \leq 180\%$ but for an α value greater than 180% the process N8 becomes relatively more attractive but not enough to disrupt the processing structure of the chemical industry. As shown in Figure 5.1, the total consumption of benzene would eventually reach the same asymptotic value as Case 2 and the total consumption of toluene reaches its maximum level of about 2.33 billion pounds at an α value greater then 1000%.

Emission Control for Acrylonitrile and Vinyl Chloride

Acrylonitrile, which is also known as vinyl cyanide, is one of the major intermediate chemicals used mainly in the manufacture of acrylic and modacrylic fibers, ABS/SAN resins, adiponitrile, nitrile rubber, and acrylamide. The National Institute for Occupational Safety and Health (NIOSH) estimates that approximately 125,000 persons are potentially exposed to this chemical in the workplace. Acrylonitrile exposure may produce nausea, vomiting, and headache. Systemic exposure to high concentrations may even cause death.

Vinyl chloride is another chemical intermediate that is regarded as a human carcinogen and a cause of angiosarcoma of the liver. Systemic exposure to this chemical depresses the central nervous system causing symptoms that resemble mild alcohol intoxication. NIOSH estimates definite worker exposure to vinyl chloride at 27,000 and probable worker exposure at 2,200,000.

It is evident that exposure to these chemicals is harmful to human health, and emission control of such materials would undoubtedly affect the performance of the certain sectors of the chemicals industry. In this section the model is used to study the effects of restrictions on the emission of acrylonitrile and vinyl chloride on the U.S. petrochemical industry. To this effect, sensitivity studies similar to those discussed earlier were conducted on the price of these two chemical intermediates.

First consider the scenario in which the price of acrylonitrile would be increased due to the restrictions on the air emission of this chemical. A factor α is again introduced to the 1985 projected industry representing the percentage increase in price of acrylonitrile. Figures 5.10 and 5.11 represent the effect of the

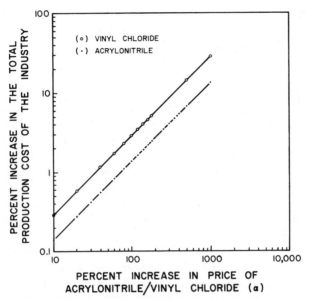

Figure 5.10 Impact of price increase for acrylonitrile and vinyl chloride on the total production cost of the industry in 1985.

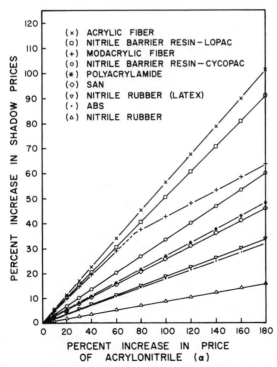

Figure 5.11 Impact of price increase for acrylonitrile on the shadow prices of polymer products in 1985.

increased values of α on the total production cost and the changes in the pricing structure of the industry, respectively. It is realized that, in general, the price increase for acrylonitrile has no significant impact on the structure of the industry. However, when α is increased to about 75% the price of acrylonitrile becomes sufficiently high to disrupt the structure of the fiber manufacturing industry. In other words, at this point ($\alpha \cong 75\%$) the batch solution polymerization process to modacrylic fiber replaces the continuous suspension process to reduce the total consumption of acrylonitrile by about 1.3%. The next structural change occurs at an α value of $380\% < \alpha \leq 390\%$. For an α value in this range, the suspension/emulsion polymerization process to ABS replaces the previously more economical bulk/suspension process. This results in about 3% further reduction in the total consumption of acrylonitrile. It is observed that at this value of α all the alternate technologies have been utilized to avoid the use of acrylonitrile and the structure of the industry remains unchanged for any α value beyond this point.

Consider next the scenario in which the price of vinyl chloride is increased systematically. Figures 5.10 and 5.12 show the impact of the higher prices for this chemical on the total production cost of the industry and the shadow prices, respectively. It is realized that no changes occur in the structure of the industry

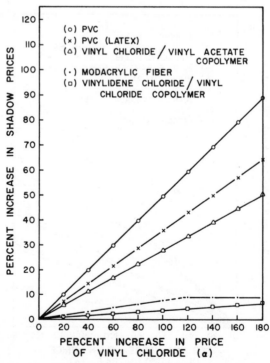

Figure 5.12 Impact of price increase for vinyl chloride on the shadow prices of polymer products in 1985.

for the values of $0\% \leq \alpha \leq 100\%$. However, at an α value over 100% (about 117%; Figure 5.12), the batch suspension polymerization process to modacrylic fiber replaces the continuous suspension process to avoid the use vinyl chloride and this results in only about 0.20% reduction in the total consumption of this chemical. The structure of the industry remains unchanged for the higher values of α.

The results of the study in this section reveal that the increased restrictions on the use of such intermediate chemicals as acrylonitrile and vinyl chloride which are closely tied to the polymer products industry have very little impact on the industry's technology selection and resource allocation. The only effect would be on the pricing structures of the polymer products as shown in Figures 5.11 and 5.12.

It, however, seems that with the current structure of the industry the possible substitutions among the end products or even penetration of natural materials into the economy would be the only solution to avoiding the use of the toxic chemicals which are closely tied to the polymer products industry.

CONCLUDING REMARKS

A systems analysis of the U.S. petrochemical industry gives insight into complex interactions between the supply and demand for intermediate chemicals used captively by the industry. Changes in technology can greatly alter pricing structures. The shadow price concept gives insight into the price changes and can provide further information upon which to base decision making.

PART TWO

Chemical Technology Catalog

The catalog includes processes for the manufacture of the major petrochemicals and is organized alphabetically by the main product in the following categories:

1   Primary feedstocks and intermediate chemicals.

2   Plastics and resins.

3   Man-made fibers.

4   Synthetic rubbers (elastomers).

5   Thermoplastic elastomers.

For each process estimates are given of the material balance, the energy requirements, and the investment. The basis of the material balance is one unit mass of the main product, with products given a positive sign and feedstocks a negative sign. The investment estimates have been adjusted to 1977 dollars, and are based on a United States Gulf Coast location.

   The utilities of electricity, steam, fuel, and so on, have been converted to *net* fuel oil equivalents which include credits for energy produced or recovered within the process. The basis of conversion (1977 $) is

1 kg of steam at 600 psig $=$ 0.79 cent

1 kg of steam at 100 psig $=$ 0.48 cent

1 kwh of electricity $=$ 1.7 cents

1 MBTU of fuel $=$ 200 cents

1 SCF of inert gas $=$ 0.04 cent

1 gal of cooling water $=$ 0.002 cent

1 gal of process water $=$ 0.027 cent

We illustrate the conversion of process data into catalog form by maleic anhydride, process 1—oxidation of benzene.

Output   1 metric ton of maleic anhydride product

Input   1.19 metric ton of benzene

Utilities

		Primary energy ($/T)
Electricity consumed	130 kwh/T	2.21
Steam produced	1.6 T/T	−7.68
Fuel consumed	7.94 MBTU/T	15.88
Inert gas consumed	353 SCF/T	0.14
Water consumed (process/cooling)		1.59
	net	12.14
		0.15 FOET*

Catalog Entry

Manufacturing Process

1   OXIDATION OF BENZENE

Material Balance

Chemical	Coefficient T/T Product
Benzene	−1.19
Maleic anhydride	1.00

Primary energy requirements for utilities   0.15 FOET/T
Unit investment for a 27 kT plant (1977 $)   0.91 $1000/T

Comments   Reaction occurs at 750°F over a vanadium pentoxide-based catalyst in a fixed bed reactor.

This information was compiled from a wide variety of sources, private and public. Permission to use proprietary information was granted with the understanding that the specific sources not be identified. Often the data from different sources conflicted: to resolve these conflicts we called upon the judg-

---

*FOET = fuel oil equivalent ton.

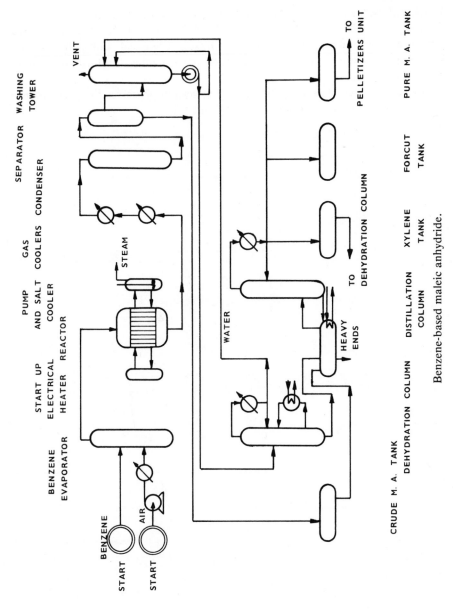

BENZENE   START UP   PUMP   SEPARATOR   WASHING
EVAPORATOR ELECTRICAL   AND SALT   TOWER
            HEATER REACTOR COOLER
                          GAS
                          COOLERS CONDENSER

Benzene-based maleic anhydride.

CRUDE M. A. TANK

DEHYDRATION COLUMN   DISTILLATION   XYLENE   FORCUT   PURE M. A. TANK
                      COLUMN   TANK   TANK

ment of those in industry familiar with the technology. A special acknowledgment is given to Dr. J. F. Mathis, Dr. D. S. Maisel, and Mr. W. W. Yuan of the Chemical Technology Department of Exxon Chemical Company. Exxon Chemical Company provided invaluable industrial contacts, information, manpower, and insight.

We have deleted from this public catalog certain processes the performance of which remains as proprietary information.

CHAPTER SIX

Primary Feedstocks and Intermediate Chemicals

ACETALDEHYDE

Acetaldehyde is used almost entirely for the synthesis of other chemicals, primarily acetic acid, acetic anhydride, n-butanol, and 2-ethylhexanol. Other routes to these chemicals are beginning to predominate however; thus the future of acetaldehyde as a large-volume chemical intermediate is somewhat clouded.

Today most acetaldehyde is produced by the catalytic liquid-phase oxidation of ethylene, the Wacker process. This process began commercial operation in 1960 and rapidly assumed a dominant position. As of 1970 a substantial fraction of the acetaldehyde produced was still derived from ethanol, which has been an important feedstock for acetaldehyde manufacture since before 1930. Ethanol can be converted to acetaldehyde by either dehydrogenation or oxidation, though the former is the usual procedure. When oxidation is used, operating conditions may be so adjusted that dehydrogenation occurs simultaneously and the exothermic heat of oxidation is balanced by the endothermic heat of dehydrogenation. Since 1945 some acetaldehyde has been manufactured by the oxidation of propane or butane. Substantial quantities of other chemicals are produced in this process; of these, formaldehyde, methanol, and acetone are the most important. The hydration of acetylene was once a major route to acetaldehyde, but by the early 1960s was no longer used in the United States.

Manufacturing Processes

1 ONE-STEP OXIDATION OF ETHYLENE

Material Balance

Chemical	Coefficient T/T Product
Acetaldehyde	1.00
Ethylene	−0.68
Oxygen	−0.40

Primary energy requirements for utilities   0.22 FOET/T
Unit Investment for a 136 kT plant (1977 $)   0.23 $1000/T

Comments Reaction occurs in liquid phase at 60 psia and 260°F with a palladium chloride catalyst promoted by cupric chloride.

2 TWO-STEP OXIDATION OF ETHYLENE

Material Balance

Chemical	Coefficient T/T Product
Acetaldehyde	1.00
Ethylene	−0.68
Hydrogen chloride	−0.02

Primary energy requirements for utilities   0.17 FOET/T
Unit investment for a 136 kT plant (1977 $)   0.32 $1000/T

Comments Reaction occurs at 220°F and 140 psia.

3 OXIDATION OF ETHANOL

Material Balance

Chemical	Coefficient T/T Product
Acetaldehyde	1.00
Ethanol	−1.20

Primary energy requirements for utilities   0.22 FOET/T

Unit investment for a 136 kT plant (1977 $)   0.17 $1000/T

ACETIC ACID

Acetic acid is used primarily for the manufacture of acetic anhydride, vinyl acetate, and various acetate esters.

Most acetic acid plants built today employ Monsanto's low pressure process, first commercialized in 1970, for the carbonylation of methanol. Another process for the carbonylation of methanol is operated, but at lower yields and much higher pressure. The oldest commercial route to acetic acid, the oxidation of acetaldehyde, is still important, accounting for about one third of acetic acid production of 1973. Another important route to acetic acid is the oxidation of butane. First commercialized in the mid-1950s, this process began to rival the acetaldehyde route by the early 1960s. Naphtha can also be oxidized to acetic acid; though this process is practiced on a large scale in Europe, it is not used in the United States. Recently there has been interest in the oxidation of butylenes to acetic acid. Two traditional sources of acetic acid, the fermentation of ethanol and the destructive distillation of wood, lost their significance many years ago, though the former is still used.

Manufacturing Processes

1 LOW PRESSURE CARBONYLATION OF METHANOL

Material Balance

Chemical	Coefficient T/T Product
Acetic acid	1.00
Carbon monoxide	−0.61
Methanol	−0.58

Primary energy requirements for utilities   0.27 FOET/T

Unit investment for a 136 kT plant (1977 $)   0.22 $1000/T

Comments   Reaction of methanol and carbon monoxide occurs at 500 psia and 390°F in presence of rhodium catalyst.

2 AIR OXIDATION OF ACETALDEHYDE

Material Balance

Chemical	Coefficient T/T Product
Acetaldehyde	−0.78
Acetic acid	1.00
Methyl acetate	0.04

Primary energy requirements for utilities   0.29 FOET/T
Unit investment for a 136 kT plant (1977 $)   0.22 $1000/T

Comments Acetaldehyde is oxidized in liquid phase at 100 psia and 140°F.
The reaction is catalyzed by magnesium acetate.

3 OXIDATION OF *N*-BUTANE

Material Balance

Chemical	Coefficient T/T Product
Acetic acid	1.00
n-Butane	−0.83
By-product credits	0.03

Primary energy requirements for utilities   0.21 FOET/T
Unit investment for a 136 kT plant (1977 $)   0.41 $1000/T

4 DIRECT OXIDATION OF *N*-BUTYLENES

Material Balance

Chemical	Coefficient T/T Product
Acetic acid	1.00
Butenes	0.09
n-Butylenes	−1.01
Formic acid	0.06
Miscellaneous chemicals	−0.01

Primary energy requirements for utilities   −0.02 FOET/T

Unit investment for a 136 kT plant (1977 $)   0.27 $1000/T

ACETIC ANHYDRIDE

Most acetic anhydride is used to manufacture cellulose acetate. Other uses include the production of aspirin and vinyl acetate.

The usual feedstocks for acetic anhydride production are ketene and acetic acid. The ketene may be obtained from either acetone or acetic acid, the latter being the usual source today. Acetic anhydride can also be produced by the oxidation of acetaldehyde and by the reaction of acetylene and acetic acid. Of these, the former retains some importance but the latter is definitely obsolete.

Manufacturing Processes

1 DIRECT OXIDATION OF ACETALDEHYDE

Material Balance

Chemical	Coefficient T/T product
Acetaldehyde	−1.08
Acetic acid	0.21
Acetic anhydride	1.00
Ethyl acetate	−0.01
By-product credits	0.03

Primary energy requirements for utilities   0.32 FOET/T

Unit investment for a 136 kT plant (1977 $)   0.42 $1000/T

Comments   Air oxidation of acetaldehyde occurs in liquid phase at 140°F and 1 atm in the presence of cobalt and copper acetate catalyst.

2 KETENE AND ACETIC ACID

Material Balance

Chemical	Coefficient T/T Product
Acetic acid	−1.25
Acetic anhydride	1.00

Primary energy requirements for utilities   0.37 FOET/T
Unit investment for a 136 kT plant (1977 $)   0.16 $1000/T

Comments   In the first step, ketene and water are formed by thermal decomposition of acetic acid in the presence of a triethyl phosphate catalyst. The vapor phase reaction is conducted at 1300°F and 100–200 mm Hg. In the second step, acetic acid and ketene are reacted in the liquid phase at 85–105°F and 50–150 mm Hg.

ACETONE

Acetone is used mainly as a solvent and as a feedstock for the manufacture of methyl methacrylate, methyl isobutyl ketone, and bisphenol-A. A variety of other chemicals, including hexylene glycol, mesityl oxide, and isophorone, are also derived from acetone.

Today about half the acetone produced is derived from isopropanol, usually by dehydrogenation. Most of the remainder occurs as a co-product of phenol manufacture from cumene. If isopropanol is oxidized to acetone, the process may be so operated that dehydrogenation occurs simultaneously, the endothermic heat of dehydrogenation balancing the exothermic heat of oxidation. In some countries (e.g., Japan) acetone is now produced by the direct oxidation of propylene in a process similar to the Wacker process for acetaldehyde production from ethylene. Ethanol, acetic acid, and acetylene have been used as feedstocks for acetone manufacture at various times, though not in the United States. A fermentation process co-producing acetone and butanol was once important, but beginning in the 1920s the process gave way to isopropanol-based processes which by 1940 represented more than 80% of acetone production.

Manufacturing Processes

1 DEHYDROGENATION OF ISOPROPANOL

Material Balance

Chemical	Coefficient T/T Product
Acetone	1.00
Isopropanol	−1.11
Nitrogen	−0.04

Primary energy requirements for utilities   0.33 FOET/T
Unit investment for a 68 kT plant (1977 $)   0.16 $1000/T

Comments   Reaction occurs in vapor phase at 750°F and 1 atm in the presence of a zinc oxide–zirconium oxide catalyst.

2 OXIDATION OF PROPYLENE

Material Balance

Chemical	Coefficient T/T Product
Acetone	1.00
Hydrogen chloride (dilute)	−0.08
Propylene (chemical grade)	−0.85
Sodium hydroxide (dilute)	−0.01
Misc. chemicals (demineralized water)	−0.43

Primary energy requirements for utilities	0.77 FOET/T
Unit investment for a 68 kT plant (1977 $)	0.33 $1000/T

ACETYLENE

Acetylene was once important as a feedstock for the production of acetal-dehyde, acrylic acid and its esters, acrylonitrile, chloroprene, perchloroethy-lene, trichloroethylene, vinyl chloride, and vinyl acetate. Some of these markets have vanished and all will vanish eventually if the current trend toward other feedstocks continues. The one growing market for acetylene is the production of 1,4-butynediol, from which a range of increasingly important chemicals is derived; these include butyrolactone, vinyl pyrrolidone, 1,4-butanediol, and tetrahydrofuran. Appreciable amounts of acetylene are also consumed in non-chemical uses such as metal cutting and welding.

The traditional route to acetylene is the hydration of coal-derived calcium carbide. As of the early 1970s this process still accounted for somewhat more than half the acetylene capacity. Since 1940 a number of processes producing acetylene by the pyrolysis of hydrocarbons have been commercialized. Essen-tially these processes differ in the way in which they supply the energy necessary for the pyrolysis. Energy may be supplied by electric discharge (arc process), by contact with a hot solid (regenerative process), by partial combustion of the feedstock (one-stage partial combustion process), or by injection of the feedstock into hot combustion gases (two-stage partial combustion process). Of these, the only process used to a significant extent in the United States has been one-stage partial combustion. There are many versions of this process, but the most widely used is the BASF or Sachsse process, which was first commercial-ized in Germany during World War II. Methane is the usual feedstock, but the process has also been adapted to use naphtha. The arc process was the first to be used for the large-scale production of acetylene. Hydrocarbons from methane to naphtha can serve as feedstock, but in practice a methane-rich gas seems to be preferred. Butane or naphtha may be used to quench the hot pyrolysis gases; in this case substantial amounts of ethylene are also produced. The only regenerative process to achieve commercial significance is the Wulff process. This was operated on a demonstration plant scale in the United States in 1951. It was not until the 1960s, however, that further commercial development occurred. Methane can serve as feedstock, but hydrocarbons from ethane to naphtha are preferred. Operating conditions can be so adjusted that substantial quantities of ethylene are co-produced. The Hoechst HTP process is the most widely used two-stage partial combustion process. It has been operated in Germany since 1960. As feedstock, any hydrocarbon from methane to naphtha can be used. Ethylene is co-produced when hydrocarbons heavier than methane are used.

Manufacturing Processes

1 SUBMERGED FLAME PROCESS

Material Balance

Chemical	Coefficient T/T Product
Acetylene	1.00
Ethylene	1.15
Fuel gas	1.30
Fuel oil (high sulfur)	−8.34
Oxygen	−7.67
Synthesis gas (H_2:CO = 2:1)	5.47

Primary energy requirements for utilities   0.04 FOET/T
Unit investment for a 136 kT plant (1977 $)   1.31 $1000/T

Comments   Reaction occurs at 130 psia and 480°F.

2 HYDRATION OF CALCIUM CARBIDE

Material Balance

Chemical	Coefficient T/T Product
Acetylene	1.00
Calcium Carbonate	−4.46
Coke	−1.86
By-product credits (waste lime)	3.22
Miscellaneous chemicals	−0.06

Primary energy requirements for utilities   2.45 FOET/T
Unit investment for a 136 kT plant (1977 $)   0.99 $1000/T

3 PYROLYSIS OF METHANE (PARTIAL OXIDATION)

Material Balance

Chemical	Coefficient T/T Product
Acetylene	1.00
Methane	−4.23
Oxygen	−4.80
Synthesis gas (H_2:CO = 2:1)	4.01

Primary energy requirements for utilities	0.57 FOET/T
Unit investment for a 136 kT plant (1977 $)	0.59 $1000/T

Comments Preheated methane and oxygen are reacted at atmospheric pressure. The combustion gases are compressed to about 11 atm to remove CO_2 by ethanolamine treatment, and pure acetylene is recovered by absorption in dimethylformamide or other solvent.

4 PYROLYSIS OF NAPHTHA (ONE-STAGE PARTIAL COMBUSTION)

Material Balance

Chemical	Coefficient T/T Product
Acetylene	1.00
Coke	−0.01
Fuel gas	0.84
Naphtha	−4.31
Oxygen	−4.31

Primary energy requirements for utilities	1.36 FOET/T
Unit investment for a 136 kT plant (1977 $)	0.74 $1000/T

5 PYROLYSIS OF ETHANE (REGENERATIVE PROCESS)

Material Balance

Chemical	Coefficient T/T Product
Acetylene	1.00
Ethane	−3.00
Fuel gas	0.57
Fuel oil (low sulfur)	0.18

Primary energy requirements for utilities	1.41 FOET/T
Unit investment for a 136 kT plant (1977 $)	0.50 $1000/T

ACROLEIN

Acrolein is used primarily for the manufacture of glycerine via allyl alcohol and for the production of acrylic acid and its esters. Of the various other chemicals derived from acrolein, methionine, a poultry feed additive, is the most important.

The first large-scale, commercial manufacture of acrolein began in the early 1950s and was based on the aldol condensation of acetaldehyde with formaldehyde. However, since the late 1950s most acrolein has been produced by oxidizing propylene.

Manufacturing Processes

1 OXIDATION OF PROPYLENE

Material Balance

Chemical	Coefficient T/T Product
Acrolein	1.00
Fuel oil (low sulfur)	0.12
Propylene (polymer grade)	−1.16

Primary energy requirements for utilities	0.92 FOET/T
Unit investment for a 18 kT plant (1977 $)	0.80 $1000/T

ACRYLAMIDE

Acrylamide is manufactured from acrylonitrile and used primarily in production of polyacrylamide which is consumed in areas such as water and effluent treatment, and adhesives.

Manufacturing Processes

1 HYDRATION OF ACRYLONITRILE (FIXED-BED CATALYST)

Material Balance

Chemical	Coefficient T/T Product
Acrylamide	1.00
Acrylonitrile	−0.76
Active carbon	−0.02
Hydrogen	−0.01

Primary energy requirements for utilities   0.38 FOET/T

Unit investment for a 14 kT plant (1977 $)   0.62 $1000/T

Comments Acrylonitrile in about 7% solution in deionized water is hydrated over an unsupported copper oxide-chromium oxide catalyst packed in a series of three reactors.

2 HYDRATION OF ACRYLONITRILE (SUSPENDED CATALYST)

Material Balance

Chemical	Coefficient T/T Product
Acrylamide	1.00
Acrylonitrile	−0.78
Active carbon	−0.02
Sodium hydroxide	−0.06
Sulfuric acid	−0.07

Primary energy requirements for utilities   0.20 FOET/T

Unit investment for a 14 kT plant (1977 $)   0.81 $1000/T

3 SULFURIC ACID PROCESS

Material Balance

Chemical	Coefficient T/T Product
Acrylamide	1.00
Acrylonitrile	−0.92
Ammonia	−0.63
Sulfuric acid	−1.77

Primary energy requirements for utilities   0.45 FOET/T

Unit investment for a 14 kT plant (1977 $)   0.91 $1000/T

ACRYLIC ACID

Most acrylic acid is consumed as ethyl acrylate or some other acrylate ester. The esterification may occur in situ if the appropriate alcohol replaces water in one of the processes described below. The acrylate esters are used to produce acrylic polymers and copolymers; these are used primarily as coatings and in textiles.

As of 1970, five different routes to acrylic acid were in commercial production: the oxidation of acrolein, the carbonylation of acetylene, the hydrolysis of acrylonitrile, the reaction of ketene and formaldehyde to beta-propiolactone for subsequent hydration, and the cyanation of ethylene oxide to form ethylene cyanohydrin for later hydrolysis. The acrolein route, commercialized here in the late 1960s, and the acetylene route, which dates back to the 1940s, are the most important, though the former is preferred today. If the acrolein route is used, a separate esterification reactor is needed when acrylate esters are desired. The ethylene oxide process, the original commercial route to acrylic acid, is today considered obsolete. The acrylonitrile and beta-propiolactone processes continue to be used; however, the former is no longer operated in the United States.

Manufacturing Processes

1 OXIDATION OF PROPYLENE (TWO-STEP VAPOR PHASE)

Material Balance

Chemical	Coefficient T/T Product
Acetic acid	0.05
Acrylic acid	1.00
Ethyl acetate	−0.01
Oxygen	−0.46
Propylene (chemical grade)	−0.72
Miscellaneous chemicals	−0.01

Primary energy requirements for utilities	0.29 FOET/T
Unit investment for a 91 kT plant (1977 $)	0.83 $1000/T

Comments   Reaction occurs at 40 psia and 617°F in the presence of a cobalt-iron-bismuth-molybdenum catalyst. Resulting acrolein is further oxidized to acrylic acid in the second-stage reactor at 30 psia and 518°F in the presence of a molybdenum-vanadium-tungsten catalyst.

2 CARBONYLATION OF ACETYLENE

Material Balance

Chemical	Coefficient T/T Product
Acetylene	−0.42
Acrylic acid	1.00
Carbon monoxide	−0.51
Methyl ethyl ketone	−0.01
Miscellaneous chemicals	−0.01

Primary energy requirements for utilities   0.58 FOET/T
Unit investment for a 45 kT plant (1977 $)   0.63 $1000/T

Comments Acetylene is compressed to 90 psia and absorbed in a tetrahydrofuran recycle stream. Reaction takes place in a carbonylation reactor which
operates at 390°F and 1200 psia.

ACRYLONITRILE

The production of acrylic and modacrylic fibers accounts for about half the demand for acrylonitrile. Most of the remainder is consumed in the production of
various copolymers, the most important of which are acrylonitrile-butadiene-
styrene, styrene-acrylonitrile, and butadiene-acrylonitrile. There is also a
substantial export market. Acrylonitrile has a potentially large-volume use as
an intermediate for producing adiponitrile, from which adipic acid and hexamethylenediamine, the monomers for nylon 66, can be derived.

Acrylonitrile was first produced commercially in the United States in 1940. In
the original manufacturing process, ethylene oxide and hydrogen cyanide
reacted to form ethylene cyanohydrin, which was then dehydrated to acrylonitrile. Though this process was operated until the mid-1960s, since the early
1950s most acrylonitrile has been manufactured from acetylene by cyanation,
or, more recently, from propylene by ammoxidation. The acetylene route was
developed during World War II but did not become important in the United
States until a decade later. The propylene route was developed in the late 1950s,
commercialized in 1960, and, because propylene was cheap and plentiful,
assumed a dominant position within a few years. Today the propylene route accounts for all the acrylonitrile produced in the United States.

Manufacturing Processes

1   AMMOXIDATION OF PROPYLENE

Material Balance

Chemical	Coefficient T/T Product
Acrylonitrile	1.00
Ammonia	−0.43
Propylene (chemical grade)	−1.20
Sulfuric acid	−0.15

Primary energy requirements for utilities   0.15 FOET/T
Unit investment for a 181 kT plant (1977 $)   0.53 $1000/T

Comments Reaction occurs in vapor phase at 760°F and 30 psia over a
bismuth-iron catalyst.

2   CYANATION/OXIDATION OF ETHYLENE

Material Balance

Chemical	Coefficient T/T Product
Acrylonitrile	1.00
Ethylene	−0.76
Hydrogen chloride	−0.17
Hydrogen cyanide	−0.60
Oxygen	−0.90

Primary energy requirements for utilities   0.60 FOET/T
Unit investment for a 181 kT plant (1977 $)   0.21 $1000/T

ACTIVE CARBON

Active carbon was used in World War I as a vapor adsorbent in military gas masks because of its ability to adsorb certain poisonous gases, and it is now widely used in both military and industrial gas masks. Active carbon is used in air conditioning systems to control odors. Industrial recovery and control of vapors is another important field of application for this material. The major use of active carbon, however, is in solution purification, such as the cleanup of cane, beet, and corn sugar solutions, and for the removal of tastes and odors from water supplies, vegetable and animal fats and oils, alcoholic beverages, chemicals, and pharmaceuticals.

Active carbon is manufactured by many carbonaceous materials, such as petroleum coke, sawdust, lignite, coal, peat, wood, charcoal, nutshells, and fruit pits, but properties of finished material are governed not only by the raw material but by the method of activation used. Sawdust and lignite are usually used for production of decolorizing active carbon. Vapor-adsorbent carbons are generally produced from coconut shells, fruit pits, and briquetted coal and charcoal.

ADIPIC ACID

Adipic acid is one of the monomers for nylon 66. The other monomer, hexamethylenediamine, can be derived from adipic acid. Most adipic acid is thus consumed directly or indirectly in nylon 66 production. Significant quantities of adipic acid are also used to produce adipate plasticizers and in the manufacture of urethane and polyester resins.

Cyclohexanol or cyclohexanone can be oxidized to adipic acid. In practice a mixture of cyclohexanol and cyclohexanone is usual feedstock. Originally the oxidation was carried out using nitric acid, but today an air oxidation process is gaining favor. Since the 1940s there have been attempts to prepare adipic acid directly from cyclohexane by oxidation. These processes have met with little commercial success.

Manufacturing Processes

1 AIR OXIDATION OF CYCLOHEXANE

Material Balance

Chemical	Coefficient T/T Product
Adipic acid	1.00
Cyclohexane	−0.78
Nitric acid, 60%	−0.90
Succinic acid	0.05
Miscellaneous Chemicals	−0.01

Primary energy requirements for utilities	0.73 FOET/T
Unit investment for a 136 kT plant (1977 $)	0.54 $1000/T

Comments Oxidation of cyclohexane takes place at low conversion (4%) using a cobalt salt catalyst. After neutralization and saponification, the resulting KA (ketone-alcohol) oil, containing about 90% cyclohexanone and cyclohexanol, is oxidized with nitric acid at 70–100°C and 3 atm, using ammonium vanadate and a copper salt as catalysts.

2 OXIDATION OF CYCLOHEXANOL WITH NITRIC ACID

Material Balance

Chemical	Coefficient T/T Product
Adipic acid	1.00
Cyclohexanol	−0.74
Nitric acid, 60%	−0.73

Primary energy requirements for utilities	0.23 FOET/T
Unit investment for a 23 kT plant (1977 $)	0.46 $1000/T

ADIPONITRILE

Essentially all adiponitrile is hydrogenated to hexamethylenediamine, one of the monomers for nylon 66.

The original process for adiponitrile manufacture used the reaction of adipic acid and ammonia. This process is still used, but has been overshadowed somewhat by a process converting butadiene to adiponitrile via dichlorobutylene and dicyanobutylene. A variation of this route starts with the chlorination of tetrahydrofuran derived from furfural, which is obtained by processing agricultural residues. In a more recently commercialized process, adiponitrile is produced by the electrolytic hydrodimerization of acrylonitrile. This development has led some to suggest the large-scale conversion of adiponitrile to adipic acid, a reversal of current practice.

Manufacturing Processes

1   REACTION OF ADIPIC ACID AND AMMONIA

Material Balance

Chemical	Coefficient T/T Product
Adipic acid	−1.37
Adiponitrile	1.00
Ammonia	−0.48
Sodium bisulfate	−0.12

Primary energy requirements for utilities	0.34 FOET/T
Unit Investment for a 23 kT plant (1977 $)	0.34 $1000/T

2   HYDRODIMERIZATION OF ACRYLONITRILE

Material Balance

Chemical	Coefficient T/T Product
Acrylonitrile	−0.96
Adiponitrile	1.00

Primary energy requirements for utilities	1.52 FOET/T
Unit investment for a 23 kT plant (1977 $)	0.57 $1000/T

ALLYL ALCOHOL

Since the late 1950s the major use of allyl alcohol has been as a feedstock for synthetic glycerine production, and the major process for its manufacture has been the reaction of acrolein and isopropanol. Before this process was introduced, allyl alcohol was derived from the hydrolysis of allyl chloride, and, to a lesser extent, from the isomerization of propylene oxide. Though these processes were still used in the 1960s neither are important today.

Manufacturing Processes

1   ISOMERIZATION OF PROPYLENE OXIDE

Material Balance

Chemical	Coefficient T/T Product
Allyl alcohol	1.00
Propylene oxide	−1.11
Miscellaneous chemicals	−0.02

Primary energy requirements for utilities	0.27 FOET/T
Unit investment for a 18 kT plant (1977 $)	0.38 $1000/T

Comments   Reaction occurs over lithium phosphate catalyst in vapor phase at 527°F and 35 psia.

2   REACTION OF ACROLEIN AND *S*-BUTANOL

Material Balance

Chemical	Coefficient T/T Product
Acrolein	−1.04
Allyl alcohol	1.00
s-butanol	−1.31
Calcium chloride	−0.08
Methyl ethyl ketone	1.24
Miscellaneous chemicals	−0.03

Primary energy requirements for utilities	0.83 FOET/T
Unit investment for a 18 kT plant (1977 $)	0.35 $1000/T

ALLYL CHLORIDE

Most allyl chloride is converted to epichlorohydrin, which is then used to manufacture epoxy resins and synthetic glycerine. The chlorination of propylene is the only large-scale commercial process for allyl chloride production; it has been in use since the late 1940s.

Manufacturing Processes

1 CHLORINATION OF PROPYLENE

Material Balance

Chemical	Coefficient T/T Product
Allyl chloride	1.00
Chlorine	−1.32
Dichloropropylenes	0.27
Hydrogen chloride	0.64
Propylene (chemical grade)	−0.73
Sodium Hydroxide	−0.01

Primary energy requirements for utilities   0.23 FOET/T

Unit investment for a 90 kT plant (1977 $)   0.22 $1000/T

Comments Reaction occurs in vapor phase at 30 psia and a maximum temperature of 950°F.

AMMONIA

Directly or indirectly, most ammonia is consumed as fertilizer. Though ammonia is primarily an agricultural chemical, large quantities are consumed by the petrochemical industry in the production of acrylonitrile, caprolactam, hexamethylenediamine, and other chemicals. Because nitric acid is manufactured from ammonia, important chemicals such as aniline and toluene diisocyanate are also ammonia derivatives.

Synthetic ammonia has long been manufactured directly from nitrogen and hydrogen. Coal-derived water-gas was once the primary source of the hydrogen required, but today this is obtained mainly from the steam reforming of hydrocarbons.

Manufacturing Processes

1   STEAM REFORMING OF NATURAL GAS

Material Balance

Chemical	Coefficient T/T Product
Ammonia	1.00
Methane	−0.42

Primary energy requirements for utilities   0.45 FOET/T
Unit investment for a 345 kT plant (1977 $)   0.23 $1000/T

Comments   Natural gas is reformed over a nickel catalyst in two stages in the presence of steam.

2   FROM NAPHTHA

Material Balance

Chemical	Coefficient T/T Product
Ammonia	1.00
Naphtha	−1.02

Primary energy requirements for utilities   0.02 FOET/T
Unit investment for a 345 kT plant (1977 $)   0.25 $1000/T

AMMONIUM SULFATE

Ammonium sulfate is produced in the manufacture of caprolactam. It is used as a fertilizer, but is considered a low-grade one compared to ammonia or ammonium nitrate. Thus, ammonium sulfate may be a waste disposal problem for caprolactam producers. Currently, however, the world-wide demand for fertilizer seems to be sufficient to absorb the co-produced ammonium sulfate.

ANILINE

Aniline is used primarily to produce antioxidants and vulcanization accelerators for rubber. Dyestuffs, once the major aniline derivatives, now represent only

10–15% of the total demand. A growing market for aniline is the production of isocyanates, in particular methylenephenylene diisocyanate.

Once manufactured largely by the reduction of nitrobenzene with iron and hydrochloric acid, aniline is now derived from nitrobenzene by catalytic hydrogenation. Some aniline has also been derived from the ammonolysis of chlorobenzene. In Japan a process producing aniline from phenol has been commercialized.

Manufacturing Processes

1 REDUCTION OF NITROBENZENE

Material Balance

Chemical	Coefficient T/T Product
Aniline	1.00
Hydrogen	−0.07
Nitrobenzene	−1.34

Primary energy requirements for utilities	0.03 FOET/T
Unit investment for a 45 kT plant (1977 $)	0.19 $1000/T

Comments Reaction occurs in vapor phase over a copper-silica catalyst at 90 psia and 520°F.

2 AMMONOLYSIS OF CYCLOHEXANOL

Material Balance

Chemical	Coefficient T/T Product
Ammonia	−0.22
Aniline	1.00
Cyclohexanol	−1.25
Cyclohexylamine	−0.13
Hydrogen	−0.06

Primary energy requirements for utilities	0.63 FOET/T
Unit investment for a 45 kT plant (1977 $)	0.25 $1000/T

3 REACTION OF PHENOL AND AMMONIA

Material Balance

Chemical	Coefficient T/T Product
Ammonia	−0.22
Aniline	1.00
Oxygen	−0.04
Phenol	−1.04
Miscellaneous Chemicals	−0.01

Primary energy requirements for utilities	0.27 FOET/T
Unit investment for a 45 kT plant (1977 $)	0.21 $1000/T

BENZENE

Though it has some solvent use, most benzene is converted to other chemicals, primarily ethylbenzene, cumene, cyclohexane, nitrobenzene, and maleic anhydride. In turn these may be converted to still other chemicals, including styrene, phenol, adipic acid, caprolactam, and aniline. Thus benzene is the starting point for a wide variety of products.

Benzene is produced in catalytic reforming operations in petroleum refineries; such operations supply most of the benzene consumed by the petrochemical industry. Coal carbonization processes, once the primary source of benzene, still account for some of the benzene supply. These supplies are augmented by production from toluene by hydrodealkylation, a process introduced in the late 1950s and now used widely. Benzene also occurs as a by-product of ethylene manufacture, particularly when naphtha or gas oil is the feedstock. In addition there are recently developed commercial processes for disproportionating toluene to benzene and xylenes.

Manufacturing Processes

1 HYDRODEALKYLATION OF TOLUENE

Material Balance

Chemical	Coefficient T/T Product
Benzene	1.00
Hydrogen	−0.07
Methane	0.24
Toluene	−1.20

Primary energy requirements for utilities   0.08 FOET/T
Unit investment for a 90 kT plant (1977 $)   0.06 $1000/T

3 DISPROPORTIONATION OF TOLUENE

Material Balance

Chemical	Coefficient T/T Product
Benzene	1.00
Fuel gas	0.01
Toluene	−2.69
Xylenes	1.61

Primary energy requirements for utilities   0.28 FOET/T
Unit investment for a 90 kT plant (1977 $)   0.09 $1000/T

BISPHENOL-A

Most bisphenol-A is consumed in the manufacture of epoxy resins. Its only other large scale use is as a monomer for polycarbonate resins. The reaction of phenol and acetone is the only commercial route to bisphenol-A.

Manufacturing Processes

1 REACTION OF PHENOL AND ACETONE

Material Balance

Chemical	Coefficient T/T Product
Acetone	−0.28
Benzene	−0.02
Bisphenol-A	1.00
Phenol	−0.89
Sodium hydroxide	−0.01
Miscellaneous chemicals	−0.01

Primary energy requirements for utilities   0.17 FOET/T
Unit investment for a 45 kT plant (1977 $)   0.58 $1000/T

Comments Reaction occurs in liquid phase at 110°F and 1 atm using anhydrous hydrogen chloride as a catalyst.

BORIC ACID

Boric acid is usually produced by acidifying a saturated solution of borax ($Na_2B_4O_7 \cdot 10H_2O$). It may also be obtained from borate minerals such as kernite ($Na_2B_4O_7 \cdot 4H_2O$), ulexite ($NaCaB_5O_9 \cdot 8H_2O$), colemanite ($Ca_2B_6O_{11} \cdot 5H_2O$), or tincal ($Na_2B_4O_7 \cdot 10H_2O$) by the same method.

The most promising market area for boric acid lies in the field of sodium-free borate products such as boric oxide and calcium borate. These products and the acid itself are used in textile glass fibers, and in reinforced plastics and belted tires.

BUTADIENE

Butadiene was not produced commercially in the United States on a large scale until the early 1940s. By 1943 huge quantities were being manufactured and then copolymerized with styrene to meet the wartime demand for the synthetic rubber GR-S (now known as SBR). This elastomer still accounts for about half the demand for butadiene. Other important derivatives are polybutadiene rubber, nitrile rubber (butadiene-acrylonitrile) and acrylonitrile-butadiene-styrene (ABS) plastic. In addition, butadiene is finding increasing use as a chemical intermediate for producing chloroprene, the monomer for neoprene rubber, and adiponitrile, an intermediate in nylon 66 manufacture.

The routes to butadiene all date back to World War II. In the United States, butadiene was produced by the dehydrogenation of n-butylenes, by the reaction of ethanol and acetaldehyde, and, to a lesser extent, by the dehydrogenation of n-butane. Some was also produced, along with ethylene and other olefins, from the thermal cracking of naphtha or gas oil. A small plant chlorinated n-butylenes and then dehydrochlorinated the chlorobutanes to butadiene. Other routes were developed in Europe. Butadiene was manufactured in Germany from acetaldehyde and from acetylene and formaldehyde. In Russia butadiene was manufactured from ethanol. In the United States today, most butadiene is derived from n-butane or n-butylenes, though increasing amounts are being made in ethylene plants using naphtha or gas oil feedstock. In Europe and Japan, ethylene plants already supply most of the butadiene.

Manufacturing Processes

1 DEHYDROGENATION OF N-BUTYLENES

Material Balance

Chemical	Coefficient T/T Product
Ammonia	−0.01
Butadiene	1.00

Chemical	Coefficient T/T Product
n-Butylenes	−1.46
Sulfuric acid	−0.01
By-product credits	0.65
Miscellaneous chemicals and catalyst	−0.01

Primary energy requirements for utilities	1.63 FOET/T
Unit investment for a 90 kT plant (1977 $)	0.74 $1000/T

2   OXIDATIVE DEHYDROGENATION OF *N*-BUTYLENES

Material Balance

Chemical	Coefficient T/T Product
Butadiene	1.00
n-Butylenes	−1.19

Primary energy requirements for utilities	1.29 FOET/T
Unit investment for a 90 kT plant (1977 $)	0.38 $1000/T

3   DEHYDROGENATION OF *N*-BUTANE

Material Balance

Chemical	Coefficient T/T Product
Butadiene	1.00
n-Butane	−1.90

Primary energy requirements for utilities	0.65 FOET/T
Unit investment for a 90 kT plant (1977 $)	0.53 $1000/T

N-BUTANE

Aside from its use as fuel, *n*-butane is important primarily as a feedstock for butadiene manufacture. It can also be used to produce ethylene, acetylene, synthesis gas, acetic acid, and other chemicals. Natural gas is the source of most of the *n*-butane consumed by the petrochemical industry; refinery gas streams account for the remainder.

1,4-BUTANEDIOL

Tetrahydrofuran (THF) is the major consumer of 1,4-butanediol. It is produced by dehydrating 1,4-butanediol and used as a solvent for polyvinyl chloride, vinylidene chloride copolymers, and polyurethanes. 1,4-Butanediol is also used in manufacture of γ-butyrolactone, polybutylene terephthalate resins, and polyurethanes.

Manufacturing Processes

1 REACTION OF ACETYLENE AND FORMALDEHYDE

Material Balance

Chemical	Coefficient T/T Product
Acetylene	−0.32
1,4-Butanediol	1.00
n-Butanol	0.03
Formaldehyde	−0.75
Hydrogen	−0.07
Miscellaneous chemicals and catalyst	−0.01

Primary energy requirements for utilities	0.56 FOET/T
Unit investment for a 27 kT plant (1977 $)	1.55 $1000/T

Comments Reaction occurs at 15 psia and 176–194°F using a copper acetylide and bismuth on silica catalyst.

2 REACTION OF BUTADIENE AND ACETIC ACID

Material Balance

Chemical	Coefficient T/T Product
Acetic acid	−0.17
Butadiene	−0.73
1,4-Butanediol	1.00
Hydrogen	−0.04
By-product credit	0.16
Miscellaneous chemicals and catalyst	−0.01

Primary energy requirements for utilities   1.29 FOET/T

Unit investment for a 27 kT plant (1977 \$)   1.65 \$1000/T

Comments   Reaction occurs at 868 psia and 176–216°F in the presence of a catalyst of palladium and tellurium on carbon.

3   REACTION OF PROPYLENE AND ACETIC ACID

Material Balance

Chemical	Coefficient T/T Product
Acetic acid	−0.04
Benzene	−0.04
1,2-Butanediol	0.15
1,4-Butanediol	1.00
Hydrogen	−0.04
Isobutanol	0.12
Methanol	−0.11
2-Methyl-1,3-propanediol	0.15
Oxygen	−0.48
Propylene (chemical grade)	−0.76
Synthesis gas (H_2:CO = 1:1)	−0.49
Miscellaneous chemicals	−0.02

Primary energy requirements for utilities   1.66 FOET/T

Unit investment for a 27 kT plant (1977 \$)   1.98 \$1000/T

4   FROM PROPYLENE OXIDE

Material Balance

Chemical	Coefficient T/T Product
Benzene	−0.02
1,4-Butanediol	1.00
Hydrogen	−0.03
2-Methyl-1,3-propanediol	0.26
Propylene oxide	−0.93
Synthesis gas (H_2:CO = 1:1)	−0.44
Miscellaneous chemicals and catalyst	−0.02

Primary energy requirements for utilities 0.31 FOET/T

Unit investment for a 27 kT plant (1977 $) 0.87 $1000/T

Comments Propylene oxide is isomerized at 527°F to allyl alcohol which is hydroformylated in the presence of a rhodium catalyst at 28 psia and 158°F and hydrogenated in the presence of a nickel catalyst at 147 psia and 104°F.

N-BUTANOL

n-Butanol is consumed largely in the production of butyl esters, primarily the acetate, acrylate, methacrylate, phthalate, and glycol esters. Substantial quantities are also used as a solvent and in the production of amine resins.

Today most *n*-butanol is manufactured from *n*-butyraldehyde derived from the oxonation of propylene. The oxo process was commercialized in Germany during World War II and in the United States in 1948. Here it has only recently overtaken the process in which *n*-butanol is produced by hydrogenating crotonaldehyde derived from acetaldehyde. A process operated commercially in Japan converts propylene and carbon monoxide directly to *n*-butanol. In the past the fermentation of carbohydrates was used to produce most *n*-butanol, but today only a small fraction is so derived.

Manufacturing Processes

1 PROPYLENE BY CONVENTIONAL OXO

Material Balance

Chemical	Coefficient T/T Product
n-Butanol	1.00
Carbon monoxide	−0.54
Fuel oil (low sulfur)	0.38
Hydrogen	−0.07
Isobutanol	0.14
Propylene (chemical grade)	−0.92

Primary energy requirements for utilities 0.43 FOET/T

Unit investment for a 68 kT plant (1977 $) 0.37 $1000/T

Comments Reaction occurs at 320°F and 4000 psia in the presence of a cobalt compound catalyst.

2 PROPYLENE USING COBALT-PHOSPHINE CATALYST

Material Balance

Chemical	Coefficient T/T Product
n-Butanol	1.00
Fuel gas	0.10
Fuel oil (low sulfur)	0.03
Isobutanol	0.11
Propylene (chemical grade)	−0.74
Synthesis gas (H_2:CO = 2:1)	−0.62
Miscellaneous chemicals and catalyst	−0.01

Primary energy requirements for utilities   0.13 FOET/T

Unit investment for a 68 kT plant (1977 $)   0.31 $1000/T

Comments   Reaction occurs in liquid phase in the presence of a soluble cobalt-phosphine catalyst at 356°F and 515 psia.

3 PROPYLENE USING RHODIUM CATALYST

Material Balance

Chemical	Coefficient T/T Product
n-Butanol	1.00
Fuel gas	0.05
Fuel oil (low sulfur)	0.03
Hydrogen	−0.03
Isobutanol	0.08
Propylene (chemical grade)	−0.69
Synthesis gas (H_2:CO = 1:1)	−0.51

Primary energy requirements for utilities   0.12 FOET/T

Unit investment for a 68 kT plant (1977 $)   0.33 $1000/T

Comments   Reaction occurs in liquid phase in the presence of a soluble rhodium-phosphine catalyst at 257°F and 100 psia.

S-BUTANOL

Though it finds some use as a solvent, most *s*-butanol is converted to methyl ethyl ketone. *s*-Butanol is manufactured from *n*-butylenes by hydration with sulfuric acid.

Manufacturing Processes

1 SULFONATION OF *N*-BUTYLENES

Material Balance

Chemical	Coefficient T/T Product
s-Butanol	1.00
n-Butylenes	−1.14
Fuel oil (low sulfur)	0.33
Sodium hydroxide (dilute)	−0.02
Sulfuric acid	−0.01

Primary energy requirements for utilities	0.71 FOET/T
Unit investment for a 45 kT plant (1977 $)	0.22 $1000/T

T-BUTANOL

t-Butanol is used as a solvent and formed by hydrolysis of *t*-butyl hydrogen sulfate, the result of reaction between isobutylene and sulfuric acid. It is also produced as a by-product of isoprene resulting from the reaction of formaldehyde and isobutylene.

BUTENE-1

Butene-1 is one of the ingredients of the refinery C_4 streams and is used on an increasing scale in copolymers with other olefins obtained by Ziegler-type techniques. Consumption of butene-1 in the manufacture of high-density polyethylene is an example of this kind. Another important application of butene-1 is in the manufacture of butylene oxide and butylene glycol. Butylene glycol is used in the production of polymeric plasticizers and is a stabilizer for chlorinated solvents.

BUTENES

Butenes (C_4 olefins) are produced during the catalytic cracking of petroleum fractions; to a lesser extent during other refinery operations; and as by-products of ethylene manufacture. By far the major use of butenes is for the production of gasoline components (alkylate and polymer gasoline), with close to 80% of all butenes being used in this application. Chemical usage of the butenes for the production of a variety of elastomers, polymers, and industrial chemicals consumes the remaining 20%.

N-BUTYLENES

Aside from its fuel use (n-butylenes are fed to refinery alkylation units), most n-butylenes are dehydrogenated to butadiene. Substantial quantities are also converted to s-butanol.

Refinery cracking units are the major supplier of n-butylenes in the United States. It is also a by-product of ethylene manufacture, and can be produced from n-butane by catalytic dehydrogenation.

N-BUTYRALDEHYDE

n-Butyraldehyde is used entirely as a chemical intermediate, primarily for n-butanol and 2-ethylhexanol. Other derivatives include butyric acid, polyvinylbutyral, and trimethylolpropane. The most important of these, butyric acid, is used primarily to produce cellulose acetate butyrate, once an important plastic in automobile manufacture.

Today most n-butyraldehyde is derived from propylene by oxonation with synthesis gas. Though this process was commercialized in Germany during World War II and in the United States in 1948, it is only within the last decade that it overtook the acetaldehyde-based process in which crotonaldehyde is hydrogenated to n-butyraldehyde. Isobutyraldehyde is produced as a by-product when the oxonation route is employed. In the past some n-butyraldehyde was manufactured by dehydrogenating n-butanol derived from the fermentation of carbohydrates.

Manufacturing Processes

1 OXONATION OF PROPYLENE

Material Balance

Chemical	Coefficient T/T Product
n-Butanol	0.05
Butyraldehyde	1.00
Carbon monoxide	−0.64
Hydrogen	−0.05
Isobutanol	0.02

Chemical	Coefficient T/T Product
Isobutyraldehyde	0.25
Propylene (chemical grade)	−0.86

Primary energy requirement for utilities	0.12 FOET/T
Unit investment for a 28 kT plant (1977 $)	0.44 $1000/T

CALCIUM CARBONATE

Calcium carbonate is a very widely used industrial chemical. It is used as a filler for artificial stone, for the neutralization of acids, and in abrasives and soaps. Substantial quantities of calcium carbonate are also used in manufacture of acetylene via calcium carbide.

CALCIUM CHLORIDE

Calcium chloride is produced in the chlorohydrin routes to ethylene and propylene oxide and in processes for manufacturing epichlorohydrin, perchloroethylene, and trichloroethylene. It is usually recovered from natural brines or as a by-product of the ammonia-soda (Solvay) process for sodium carbonate. Its main use is in deicing and dust control on roads or sidewalks. It is also used in cement mixes to hasten setting.

CALCIUM OXIDE

Calcium oxide is produced from calcium carbonate (limestone) and is used in a variety of different applications such as insecticides, plant and animal food, water softening, and medicinal purposes. It is also consumed by the petrochemical industry in manufacture of ethylene oxide, propylene oxide, glycerine, and epicholorhydrin.

CAPROLACTAM

Caprolactam is the monomer for nylon 6; this is its only outlet. Most nylon 6 is used as fiber for carpets, textiles, or tires. Nylon 6 was first produced in Europe in the 1940s but did not come to the United States until the mid-1950s.

In the usual route to caprolactam, cyclohexanone reacts with hydroxylamine sulfate to form cyclohexanone oxime, which then undergoes rearrangements. Large quantities of ammonium sulfate, a low-grade fertilizer, are co-produced in the process, and can be a disposal problem, though this is apparently not the case today. Much research has been devoted to developing processes which reduce or eliminate the coproduction of ammonium sulfate. Three such processes were commercialized in the 1960s. In the United States, processes involving the nitration of cyclohexane with nitric acid and the reaction of cyclohexanone with peracetic acid were used but have since been shut down. In Japan a process in which cyclohexane is reacted with nitrosyl chloride is used. Another route to caprolactam uses toluene-derived benzoic acid as the feedstock; this is operated in Italy. This process offers no significant reduction in ammonium sulfate production.

Manufacturing Processes

1   VIA HEXAHYDROBENZOIC ACID

Material Balance

Chemical	Coefficient T/T Product
Ammonia	−1.32
Ammonium sulfate	4.25
Caprolactam	1.00
Fuel oil (low sulfur)	0.05
Hydrogen	−0.08
Oleum	−3.15
Sodium hydroxide	−0.30
Toluene	−1.11

Primary energy requirements for utilities	1.03 FOET/T
Unit investment for a 68 kT plant (1977 $)	1.50 $1000/T

2   NITRIC OXIDE REDUCTION PROCESS

Material Balance

Chemical	Coefficient T/T Product
Ammonia	−0.92
Ammonium sulfate	2.60
Caprolactam	1.00
Cyclohexane	−1.06
Hydrogen	−0.04
Oleum	−2.10
Oxygen	−0.55
By-product credit	0.04

Primary energy requirements for utilities   1.26 FOET/T
Unit investment for a 68 kT plant (1977 $)   1.65 $1000/T

Comments   Reaction occurs in liquid phase.

3   PHENOL PROCESS

Material Balance

Chemical	Coefficient T/T Product
Ammonia	−1.48
Ammonium sulfate	4.40
Caprolactam	1.00
Hydrogen	−0.05
Oleum	−1.36
Phenol	−0.92
Sulfur	−0.67
By-product credit	0.05

Primary energy requirements for utilities   0.65 FOET/T
Unit investment for a 68 kT plant (1977 $)   1.27 $1000/T

Comments   Reaction occurs at about 365°F and 185 psia using a Pd-on-carbon catalyst.

4   PHOTONITROSATION OF CYCLOHEXANE

Material Balance

Chemical	Coefficient T/T Product
Ammonia	−0.68
Ammonium sulfate	1.85
Caprolactam	1.00
Cyclohexane	−0.91
Hydrogen chloride	−0.05
Oleum	−1.39

Primary energy requirements for utilities   1.78 FOET/T
Unit investment for a 68 kT plant (1977 $)   1.60 $1000/T

5   CYCLOHEXANONE AND HYDROXYLAMINE

Material Balance

Chemical	Coefficient T/T Product
Ammonia	−0.93
Ammonium sulfate	2.96

Chemical	Coefficient T/T Product
Caprolactam	1.00
Cyclohexanone	−0.96
Fuel oil (low sulfur)	0.07
Hydrogen	−0.04
Oleum	−2.76
Oxygen	−0.49
Sulfuric acid	−0.93
By-product credit	0.10

Primary energy requirements for utilities	0.20 FOET/T
Unit investment for a 68 kT plant (1977 $)	1.24 $1000/T

CARBON DIOXIDE

Carbon dioxide is produced from carbon monoxide when the water-gas shift reaction is used to remove the carbon monoxide from the synthesis gas for ammonia production. Frequently the ammonia manufacturer uses this carbon dioxide to produce urea. The off-gases from various other petrochemical processes also contain carbon dioxide. The production of dry ice and carbonated beverages account for most of the demand for carbon dioxide. The amounts so consumed are produced primarily by the combustion of fuels. Carbon dioxide is also produced from limestone, recovered from fermentation processes, and drawn from natural wells.

CARBON MONOXIDE

Carbon monoxide is produced along with hydrogen in the steam reforming or partial oxidation of hydrocarbons. It is used almost entirely as a feedstock for producing methanol, acetic acid, n-butyraldehyde, and other chemicals.

Manufacturing Processes

1 STEAM REFORMING OF NATURAL GAS

Material Balance

Chemical	Coefficient T/T Product
Carbon monoxide	1.00
Hydrogen	0.23
Methane	−0.64

Primary energy requirements for utilities   0.52 FOET/T
Unit investment for a 159 kT plant (1977 \$)   0.17 \$1000/T

Comments Reaction occurs at 1400–1600°F and 150–350 psia over a nickel oxide catalyst.

2 FROM NAPHTHA

Material Balance

Chemical	Coefficient T/T Product
Carbon monoxide	1.00
Hydrogen	0.25
Naphtha	−0.80

Primary energy requirements for utilities   0.06 FOET/T
Unit investment for a 159 kT plant (1977 \$)   0.17 \$1000/T

CHLORINE

Chlorine is consumed in the production of several large-volume petrochemicals, including vinyl chloride, trichloroethylene, perchloroethylene, and carbon tetrachloride. In addition there are chemicals such as propylene oxide, ethylene oxide, and adiponitrile, which contain no chlorine, but which are or have been manufactured using processes requiring chlorine. The petrochemical industry is the largest single outlet for chlorine.

Most chlorine is derived from the electrolysis of sodium chloride. Sodium chloride generated as a co-product in the petrochemical industry can be recycled to produce additional chlorine.

Manufacturing Processes

1 ELECTROLYSIS OF SODIUM CHLORIDE IN DIAPHRAGM-CELLS

Material Balance

Chemical	Coefficient T/T Product
Chlorine	1.00
Hydrogen (as fuel gas)	0.03
Sodium carbonate	−0.04
Sodium chloride	−1.68
Sodium hydroxide	1.12
Sulfuric acid	−0.08

Primary energy requirements for utilities   0.84 FOET/T
Unit investment for a 181 kT plant (1977 $)   0.40 $1000/T

CHLOROBENZENE

Historically the major use for chlorobenzene has been in the manufacture of phenol. However, the processes producing phenol from chlorobenzene have now largely been replaced; thus the production of chlorobenzene has declined in recent years. Restrictions on the use of DDT, another major chlorobenzene derivative, have contributed to this decline. Other outlets for chlorobenzene include the synthesis of intermediates for the production of dyes and pesticides.

Chlorobenzene can be produced by the chlorination or oxychlorination of benzene. Both processes have been used on a very large scale, though the oxychlorination process was used only in connection with phenol production and has now been abandoned.

Manufacturing Processes

1 CHLORINATION OF BENZENE

Material Balance

Chemical	Coefficient T/T Product
Benzene	−0.77
Chlorine	−0.79
Chlorobenzene	1.00
Dichlorobenzene	0.11
Hydrogen chloride (dilute)	0.40

Primary energy requirements for utilities   0.12 FOET/T
Unit investment for a 55 kT plant (1977 $)   0.12 $1000/T

2 OXYCHLORINATION OF BENZENE

Material Balance

Chemical	Coefficient T/T Product
Benzene	−0.78
Chlorobenzene	1.00
Hydrogen chloride	−0.50
Sodium hydroxide	−0.01

Primary energy requirements for utilities   0.20 FOET/T
Unit investment for a 55 kT plant (1977 \$)   0.12 \$1000/T

CHLOROPRENE

Essentially all chloroprene is converted to polychloroprene (neoprene), the first commercially successful synthetic rubber. From 1932, when production began, until 1970, nearly all chloroprene produced in the United States was derived from acetylene. Today, however, this route is virtually obsolete, chloroprene now being derived from butadiene. The butadiene route was commercialized in France, in the early 1960s; once it was adopted in the United States the swing away from the acetylene route was rapid.

Manufacturing Processes

1   CHLORINATION OF BUTADIENE

Material Balance

Chemical	Coefficient T/T Product
Butadiene	−0.71
Chlorine	−0.89
Chloroprene	1.00
Sodium hydroxide	−0.66
Miscellaneous chemicals	−0.02

Primary energy requirements for utilities   0.19 FOET/T
Unit investment for a 45 kT plant (1977 \$)   0.32 \$1000/T

Comments   Chlorination of butadiene occurs at 640°F and 21 psia to form 1,4 and 3,4-dichlorobutene (DCB) isomers. The 1,4-DCB is then catalytically isomerized in presence of $CuCl_2$ and alphapicoline to provide 3,4-DCB. The 3,4-DCB is then dehydrochlorinated at 195°F and 40 psia in presence of caustic to form chloroprene.

2   DIMERIZATION OF ACETYLENE

Material Balance

Chemical	Coefficient T/T Product
Acetylene	−0.68
Chloroprene	1.00
Hexane	−0.01
Hydrogen chloride	−0.49

Primary energy requirements for utilities   0.26 FOET/T

Unit investment for a 45 kT plant (1977 $)   0.34 $1000/T

COAL

Coal is the most important of the solid fuels which is consumed in steam-electrical plants. The growing scarcity of crude oil and natural gas in the United States has directed attention to coal as a supplier of carbon atoms necessary for petrochemicals. The coal is converted to synthesis gas ($CO + H_2$) which is subsequently converted to methanol or by Fischer-Tropsch technology to liquid hydrocarbons.

COKE

Coke is the porous, solid residue resulting from the incomplete combustion of coal heated in a closed chamber. Substantial quantities of coke are consumed in steel industry and also in producing acetylene via calcium carbide.

CUMENE

Today cumene is used almost entirely for phenol production. During World War II large quantities of cumene were manufactured for use in aviation gasoline. All cumene is produced by alkylating benzene with propylene.

Manufacturing Processes

1 REACTION OF BENZENE AND PROPYLENE

Material Balance

Chemical	Coefficient T/T Product
Benzene	−0.67
Cumene	1.00
Propylene (chemical grade)	−0.38
By-product credit	0.05

Primary energy requirements for utilities   0.06 FOET/T

Unit investment for a 127 kT plant (1977 $)   0.12 $1000/T

Comments   Reaction occurs at 440°F and 515 psia over a phosphoric-acid-on-kieselguhr catalyst.

CYCLOHEXANE

Most cyclohexane is oxidized to cyclohexanol and cyclohexanone, which are subsequently converted to adipic acid for nylon 66 manufacture and caprolactam for nylon 6 manufacture. Substantial quantities of cyclohexane are exported. Cyclohexane is manufactured by the hydrogenation of benzene. In addition there is a supply of cyclohexane obtained directly from petroleum by fractionation.

Manufacturing Processes

1 HYDROGENATION OF BENZENE

Material Balance

Chemical	Coefficient T/T Product
Benzene	−0.94
Cyclohexane	1.00
Fuel gas	0.03
Hydrogen	−0.07

Primary energy requirements for utilities	−0.01 FOET/T
Unit investment for a 100 kT plant (1977 $)	0.08 $1000/T

Comments   The vapor phase reaction occurs in two steps using a nickel catalyst. The first reaction occurs at 428°F and 32 atm. The reaction is then completed in the adiabatic second stage reactor.

CYCLOHEXANOL

Cyclohexanol is used almost entirely for the production of adipic acid and caprolactam for eventual consumption as nylon 6. A mixture of cyclohexanol and cyclohexanone is suitable for adipic acid manufacture. If used for caprolactam production, cyclohexanol must first be dehydrogenated to cyclohexanone.

   Most cyclohexanol is derived from cyclohexane by oxidation. Cyclohexanone is co-produced in this process. In a fairly recent development, this oxidation is carried out in the presence of boric acid, thereby increasing the overall yield of cyclohexanol and cyclohexanone. Cyclohexanol was originally derived from phenol. Although this process gave way to the cyclohexane route in the 1940s, it is still used and cannot be regarded as obsolete.

Manufacturing Processes

1 OXIDATION OF CYCLOHEXANE

Material Balance

Chemical	Coefficient T/T Product
Cyclohexane	−1.64
Cyclohexanol	1.00
Cyclohexanone	0.38
Sodium hydroxide	−0.13

Primary energy requirements for utilities	0.43 FOET/T
Unit investment for a 23 kT plant (1977 $)	0.55 $1000/T

2 OXIDATION OF CYCLOHEXANE (BORON-ASSISTED)

Material Balance

Chemical	Coefficient T/T Product
Boric acid	−0.01
Cyclohexane	−1.00
Cyclohexanol	1.00
Cyclohexanone	0.07
Sodium hydroxide	−0.35

Primary energy requirements for utilities	0.56 FOET/T
Unit investment for a 23 kT plant (1977 $)	0.44 $1000/T

3 OXIDATION OF CYCLOHEXANE

Material Balance

Chemical	Coefficient T/T Product
Boric acid	−0.01
Cyclohexane	−0.91
Cyclohexanol	1.00

Primary energy requirements for utilities	0.83 FOET/T
Unit investment for a 23 kT plant (1977 $)	0.51 $1000/T

CYCLOHEXANONE

Virtually all cyclohexanone is used to produce adipic acid and caprolactam, both of which are monomers for nylon manufacture. A mixture of cyclohexanone and cyclohexanol is suitable for adipic acid production.

Most cyclohexanone is obtained as a co-product with cyclohexanol from the oxidation of cyclohexane. Significant quantities are also produced from cyclohexanol by dehydrogenation and from phenol by hydrogenation. These two routes account for much of the cyclohexanone converted to caprolactam.

Manufacturing Processes

1 DEHYDROGENATION OF CYCLOHEXANOL

Material Balance

Chemical	Coefficient T/T Product
Cyclohexanol	−1.09
Cyclohexanone	1.00

Primary energy requirements for utilities	0.09 FOET/T
Unit investment for a 68 kT plant (1977 $)	0.12 $1000/T

2 OXIDATION/DEHYDROGENATION OF CYCLOHEXANE

Material Balance

Chemical	Coefficient T/T Product
Cyclohexane	−1.05
Cyclohexanone	1.00
Hydrogen (as fuel gas)	0.02

Primary energy requirements for utilities	0.34 FOET/T
Unit investment for a 68 kT plant (1977 $)	0.46 $1000/T

DICHLOROBENZENE

Dichlorobenzene is produced with monochlorobenzene in the benzene chlorination process. It is formed in two isomeric forms, ortho and para, but the para compound is formed in the larger quantity with the ratio of para to ortho of approximately 3:1. The composition of the chlorinated products, however, varies according to the chlorination temperature, rate, and catalyst.

o-Dichlorobenzene is used mainly for pesticides. It is also consumed as a solvent in paint removers and engine cleaners, and as a solvent carrier in the manufacture of toluene diisocyanate. p-Dichlorobenzene is used extensively as a moth repellent because of its pleasant odor.

DIETHYLENE GLYCOL

The major use for diethylene glycol is the production of polyurethane and unsaturated polyester resins. Other important outlets are triethylene glycol production, natural gas dehydration, textile dyeing, solvent extraction, and plasticizer and surfactant manufacture.

Diethylene glycol is produced as a by-product when ethylene oxide is hydrated to ethylene glycol. The quantities so produced have been sufficient to satisfy the demand. Should additional diethylene glycol be required it can be produced by reacting ethylene oxide and ethylene glycol. It is this reaction that is responsible for the formation of by-product diethylene glycol; the extent to which it occurs can be increased by decreasing the water to ethylene oxide ratio in the feed to the hydration process.

DIMETHYL TEREPHTHALATE

Dimethyl terephthalate (DMT) is used almost entirely for the production of polyethylene terephthalate, a polyester first prepared commercially in the United States in the early 1950s. Until the mid-1960s all polyethylene terephthalate was manufactured from DMT and ethylene. As the purification processes for terephthalic acid (TPA) become more economically feasible, this chemical is gradually displacing DMT as the preferred monomer for polyethylene terephthalate production. DMT was first prepared commercially by esterifying terephthalic acid. Later a stepwise oxidation and esterification starting with p-xylene was developed. As of the early 1970s more DMT was manufactured via the direct esterification route, though the alternative was also operated on a very large scale.

Manufacturing Processes

1 SUCCESSIVE OXIDATION AND ESTERIFICATION OF P-XYLENE

Material Balance

Chemical	Coefficient T/T Product
Dimethyl terephthalate	1.00
Methanol	−0.41
p-Xylene	−0.63

Primary energy requirements for utilities   0.32 FOET/T
Unit investment for a 150 kT plant (1977 $)   0.82 $1000/T

Comments The oxidation of *p*-xylene occurs at 302°F and 103 psia in the presence of a cobalt-manganese catalyst. The esterification reactor operates at 518°F and 397 psia.

2 ESTERIFICATION OF TEREPHTHALIC ACID

Material Balance

Chemical	Coefficient T/T Product
Dimethyl terephthalate	1.00
Methanol	−0.35
Terephthalic acid (crude)	−0.88

Primary energy requirements for utilities   0.40 FOET/T
Unit investment for a 150 kT plant (1977 $)   0.30 $1000/T

DINITROTOLUENE

The only major market for dinitrotoluene is the production of toluene diamine for conversion to toluene diisocyanate. The latter is used to produce polyurethane plastics. Dinitrotoluene is manufactured by nitrating toluene with nitric acid. An 80:20 mixture of the 2,4 and 2,6 isomers is the usual commercial product.

Manufacturing Processes

1 NITRATION OF TOLUENE

Material Balance

Chemical	Coefficient T/T Product
Dinitrotoluene	1.00
Nitric acid, 60%	−0.18
Nitric acid, 95%	−0.54
Sulfuric acid	−0.03
Toluene	−0.53

Primary energy requirements for utilities   0.12 FOET/T
Unit investment for a 55 kT plant (1977 $)   0.22 $1000/T

DIPROPYLENE GLYCOL

The most important use of diproplene glycol is as a nontoxic plasticizer for cellophane. It is also consumed in combination with diethylene glycol in the 'Udex' aromatics extraction process. Dipropylene glycol is a co-product with propylene glycol from the propylene oxide hydration process.

EPICHLOROHYDRIN

Once important only as an intermediate for synthetic glycerine production, epichlorohydrin is now also important as a raw material for the manufacture of epoxy resins. These two uses account for nearly all epichlorohydrin produced.

Since 1948, when it was first produced on a large scale, most epichlorohydrin has been derived from allyl chloride. The conversion of allyl alcohol to epichlorohydrin has also been practiced commercially.

Manufacturing Processes

1   CHLOROHYDRINATION OF ALLYL CHLORIDE

Material Balance

Chemical	Coefficient T/T Product
Allyl chloride	−0.98
Calcium oxide	−0.76
Chlorine	−0.90
Epichlorohydrin	1.00

Primary energy requirements for utilities	0.40 FOET/T
Unit investment for a 90 kT plant (1977 $)	0.18 $1000/T

Comments   Reaction is conducted adiabatically in liquid phase at 20 psia and 125°F.

ETHANE

Ethane is important primarily as a feedstock for ethylene manufacture, though it can also be used to produce acetylene, synthesis gas, and chlorinated solvents such as methyl chloroform and perchloroethylene. Natural gas is the source of most of the ethane consumed by the petrochemical industry; refinery gas streams account for the remainder.

ETHANOL

Ethanol is a versatile solvent and chemical intermediate. As a solvent it is used primarily in surface coatings, hair and scalp preparations, cosmetics, and pharmaceuticals. Ethanol was once used primarily as an intermediate for acetaldehyde, but this is no longer a major market as ethylene is now converted directly to acetaldehyde. Ethyl acetate, ethyl acrylate, ethylene glycol monoethyl ether, and other esters and ethers account for most of the demand for ethanol as a chemical intermediate.

Today most ethanol is produced by the direct catalytic hydration of ethylene, a process first commercialized in the latter 1940s. Until the mid-1960s, however, most synthetic ethanol was produced from ethylene via ethyl sulfate. Before 1930 all industrial ethanol was manufactured by fermentation. Although some is still so produced, since the mid-1940s most has been derived synthetically from ethylene.

Manufacturing Processes

1 HYDRATION OF ETHYLENE

Material Balance

Chemical	Coefficient T/T Product
Ethanol	1.00
Ethylene	−0.75
Fuel gas	0.06

Primary energy requirements for utilities   0.74 FOET/T
Unit investment for a 272 kT plant (1977 $)   0.39 $1000/T

Comments   Reaction occurs at 1000 psia and 540°F over a catalyst of phosphoric acid on a diatomaceous earth support.

ETHYL ACETATE

Ethyl acetate is used almost entirely as a solvent, primarily for coatings. It is produced commercially from acetic acid by esterification with ethanol. Another process used commercially employs the Tischtschenko reaction to produce ethyl acetate from acetaldehyde.

ETHYL ACRYLATE

Ethyl acrylate and other acrylate esters are used almost entirely to produce acrylic polymers and copolymers. These are used primarily as coatings and in the textile industry. Ethyl acrylate is produced from acrylic acid by esterification. This may be done in situ by using ethanol in place of water in one of the processes for acrylic acid manufacture.

Manufacturing Processes

1 ESTERIFICATION OF ACRYLIC ACID

Material Balance

Chemical	Coefficient T/T Product
Acrylic acid	−0.77
Ethanol	−0.48
Ethyl acrylate	1.00
Sodium hydroxide	−0.07
Sulfuric acid	−0.06

Primary energy requirements for utilities	0.39 FOET/T
Unit investment for a 45 kT plant (1977 $)	0.30 $1000/T

2 HYDROLYSIS/ESTERIFICATION OF ACRYLONITRILE

Material Balance

Chemical	Coefficient T/T Product
Acrylonitrile	−0.58
Ethanol	−0.53
Ethyl acrylate	1.00
Sulfuric acid	−1.20
Miscellaneous chemicals	−0.01

Primary energy requirements for utilities	1.13 FOET/T
Unit investment for a 23 kT plant (1977 $)	0.44 $1000/T

3   CARBONYLATION OF ACETYLENE

Material Balance

Chemical	Coefficient T/T Product
Acetylene	−0.32
Carbon monoxide	−0.35
Ethanol	−0.53
Ethyl acrylate	1.00
Hydrogen chloride	−0.04
Sodium carbonate	−0.04
Sodium hydroxide (dilute)	−0.10
Miscellaneous chemicals and catalyst	−0.02

Primary energy requirements for utilities   0.47 FOET/T
Unit investment for a 23 kT plant (1977 $)   0.41 $1000/T

ETHYLBENZENE

Styrene production accounts for essentially all the demand for ethylbenzene, most of which is manufactured from benzene and ethylene. Since 1957 some ethylbenzene has been recovered from refinery reformate streams. Today this accounts for about 10% of ethylbenzene production.

Manufacturing Processes

1   ALKYLATION OF BENZENE

Material Balance

Chemical	Coefficient T/T Product
Benzene	−0.74
Ethylbenzene	1.00
Ethylene	−0.27
Fuel oil (high sulfur)	0.01

Primary energy requirements for utilities   0.02 FOET/T
Unit investment for a 522 kT plant (1977 $)   0.06 $1000/T

Comments   Reaction occurs in liquid phase at about 320°F and 7.8 atm in presence of aluminum trichloride catalyst with HCl promoter.

ETHYL CHLORIDE

Most ethyl chloride is converted to tetraethyl lead for gasoline antiknock fluid. Since most cars manufactured after 1974 require unleaded gasoline, the demand for ethyl chloride can be expected to decline. Ethyl chloride is also used to manufacture ethyl cellulose plastics, and various dyes and pharmaceuticals. In addition it has some use as a solvent and as a refrigerant. For many years ethyl chloride was manufactured from ethanol and hydrogen chloride. By the mid-1940s, however, most ethyl chloride was being derived from ethylene and hydrogen chloride; this remains the most important route. Some ethyl chloride is also produced by chlorinating ethane.

ETHYLENE

Next to ammonia, ethylene is the petrochemical produced in the largest tonnage. Over half the ethylene produced in the United States is converted to other chemicals, primarily acetaldehyde, ethanol, ethylene oxide, ethyl chloride, ethylene dichloride, ethylbenzene, and vinyl acetate. From these chemicals is produced a wide range of other chemicals, including acetic acid, ethylene glycol, styrene, and vinyl chloride. The largest single market for ethylene is the manufacture of polyethylene. Ethylene is also polymerized with propylene to yield elastomers.

Ethylene is manufactured by pyrolysis from ethane, propane, *n*-butane, naphtha, or gas oil. In the United States the usual feedstocks are ethane and propane obtained from refinery gas streams or natural gas. Ethylene itself is present in refinery gases but not at a concentration sufficient to make large-scale recovery attractive. In Europe and Japan, where natural gas is not rich in ethane or propane, the usual feedstocks are naphtha and gas oil. The use of heavier feedstocks is increasing gradually in the United States. The pyrolysis is carried out in the presence of steam in tubular furnace coils in an operation known as "coil cracking" or "steam cracking." Various other means of pyrolysis, some of which can accept crude oil as the feedstock, have been developed but have not been used to a significant extent. Many of the pyrolysis operations for producing acetylene also yield ethylene.

Alternatives to hydrocarbon pyrolysis for producing ethylene are ethanol dehydration and acetylene hydrogenation. Neither process has been important in the United States, but both have been used extensively abroad.

Manufacturing Processes

1 STEAM CRACKING OF ETHANE-PROPANE (50-50 WT.%)

Material Balance

Chemical	Coefficient T/T Product
Butenes	0.06
Ethane	−0.92
Ethylene	1.00
Fuel gas	0.48
Propane	−0.92
Propylene (chemical grade)	0.14
Pyrolysis gasoline	0.16

Primary energy requirements for utilities	0.87 FOET/T
Unit investment for a 454 kT plant (1977 $)	0.36 $1000/T

2 STEAM CRACKING OF GAS OIL (HIGH SEVERITY)

Material Balance

Chemical	Coefficient T/T Product
Butenes	0.46
Ethylene	1.00
Fuel oil (low sulfur)	1.09
Gas oil	−4.50
Propylene (chemical grade)	0.70
Pyrolysis gasoline	0.74

Primary energy requirements for utilities	1.20 FOET/T
Unit investment for a 454 kT plant (1977 $)	0.64 $1000/T

3 STEAM CRACKING OF NAPHTHA (HIGH SEVERITY)

Material Balance

Chemical	Coefficient T/T Product
Butenes	0.37
Ethylene	1.00

Chemical	Coefficient T/T Product
Fuel gas	0.58
Fuel oil (low sulfur)	0.05
Naphtha	−3.25
Propylene (chemical grade)	0.63
Pyrolysis gasoline	0.62

Primary energy requirements for utilities   0.94 FOET/T
Unit investment for a 454 kT plant (1977 $) $1000/T

4 PYROLYSIS OF ETHANE

Material Balance

Chemical	Coefficient T/T Product
Butenes	0.04
Ethane	−1.30
Ethylene	1.00
Fuel gas	0.18
Propylene (chemical grade)	0.04
Pyrolysis gasoline	0.05
Sodium hydroxide	−0.01

Primary energy requirements for utilities   0.74 FOET/T
Unit investment for a 454 kT plant (1977 $)   0.33 $1000/T

5 PYROLYSIS OF PROPANE

Material Balance

Chemical	Coefficient T/T Product
Butenes	0.08
Ethylene	1.00
Fuel gas	0.77
Propane	−2.36
Propylene (chemical grade)	0.24
Pyrolysis gasoline	0.27

Primary energy requirements for utilities   1.01 FOET/T
Unit investment for a 454 kT plant (1977 $)   0.40 $1000/T

6 PYROLYSIS OF NAPHTHA (LOW SEVERITY)

Material Balance

Chemical	Coefficient T/T Product
Butenes	0.36
Ethylene	1.00
Fuel gas	0.50
Fuel oil (low sulfur)	0.25
Naphtha	−3.92
Propylene (chemical grade)	0.60
Pyrolysis gasoline	1.21

Primary energy requirements for utilities   0.93 FOET/T
Unit investment for a 454 kT plant (1977 $)   0.62 $1000/T

7 PYROLYSIS OF GAS OIL (LOW SEVERITY)

Material Balance

Chemical	Coefficient T/T Product
Butenes	0.53
Ethylene	1.00
Fuel gas	0.39
Fuel oil (low sulfur)	2.16
Gas oil	−6.02
Propylene (chemical grade)	0.85
Pyrolysis gasoline	1.10

Primary energy requirements for utilities   1.25 FOET/T
Unit investment for a 454 kT plant (1977 $)   0.66 $1000/T

8 STEAM CRACKING OF GAS OIL (MEDIUM SEVERITY)

Material Balance

Chemical	Coefficient T/T Product
Butenes	0.57
Ethylene	1.00
Fuel oil (low sulfur)	1.25
Gas oil	−5.53
Propylene (chemical grade)	0.80
Pyrolysis gasoline	1.40

Primary energy requirements for utilities   1.25 FOET/T
Unit investment for a 454 kT plant (1977 \$)   0.67 \$1000/T

9   HYDROGENATION OF ACETYLENE

Material Balance

Chemical	Coefficient T/T Product
Acetylene	−1.09
Ethylene	1.00
Fuel gas	1.24
Hydrogen	−0.31

Primary energy requirements for utilities   0.21 FOET/T
Unit investment for a 23 kT plant (1977 \$)   0.14 \$1000/T

10   DEHYDRATION OF ETHANOL

Material Balance

Chemical	Coefficient T/T Product
Ethanol	−1.75
Ethylene	1.00

Primary energy requirements for utilities   0.20 FOET/T
Unit investment for a 454 kT plant (1977 \$)   0.14 \$1000/T

ETHYLENE DICHLORIDE

Most ethylene dichloride is converted to vinyl chloride. The production of chlorinated solvents such as trichloroethylene, perchloroethylene, and methyl chloroform represents the next largest market. Other important derivatives are vinylidene chloride and ethylenediamine. Ethylene dichloride is also used with ethylene dibromide as a lead scavenger in leaded gasoline.

Ethylene can be chlorinated or oxychlorinated to ethylene dichloride. The oxychlorination route was developed in the 1950s as an outlet for the hydrogen chloride produced as a by-product in the conversion of ethylene dichloride to vinyl chloride. Today most manufacturers operate both chlorination and oxychlorination processes.

Manufacturing Processes

1 CHLORINATION OF ETHYLENE

Material Balance

Chemical	Coefficient T/T Product
Chlorine	−0.70
Ethylene	−0.36
Ethylene dichloride	1.00

Primary energy requirements for utilities   0.10 FOET/T
Unit investment for a 272 kT plant (1977 $)   0.02 $1000/T

2 OXYCHLORINATION OF ETHYLENE

Material Balance

Chemical	Coefficient T/T Product
Ethylene	−0.32
Ethylene dichloride	1.00
Hydrogen chloride (dilute)	−0.94

Primary energy requirements for utilities   0.01 FOET/T
Unit investment for a 272 kT plant (1977 $)   0.04 $1000/T

ETHYLENE GLYCOL

Ethylene glycol is used as an antifreeze in automotive cooling systems, and is the monomer for ethylene glycol terephthalate, a polyester. These are the two major markets for ethylene glycol. It is also used in deicing compounds, in latex paints and other emulsions, in brake and shock absorber fluid, as a solvent for certain dyes, for the production of explosives, and for various other purposes.

The hydration of ethylene oxide is the only commercial route to ethylene glycol used today. This may soon change, because promising new technologies producing ethylene glycol directly from ethylene or synthesis gas are nearing commercialization. Until the late 1960s some ethylene glycol was derived from formaldehyde and synthesis gas.

Manufacturing Processes

1 HYDRATION OF ETHYLENE OXIDE

Material Balance

Chemical	Coefficient T/T Product
Diethylene glycol	0.11
Ethylene glycol	1.00
Ethylene oxide	−0.87
Triethylene glycol	0.03

Primary energy requirements for utilities   0.33 FOET/T

Unit investment for a 181 kT plant (1977 $)   0.15 $1000/T

Comments   The noncatalytic liquid phase reaction occurs at 392°F and 215 psia.

2 OXIDATION OF ETHYLENE

Material Balance

Chemical	Coefficient T/T Product
Acetic acid	−0.03
Ethylene	−0.51
Ethylene glycol	1.00
Methane	−0.02
Oxygen	−0.31
Miscellaneous chemicals	−0.01

Primary energy requirements for utilities   0.64 FOET/T

Unit investment for a 181 kT plant (1977 $)   0.39 $1000/T

Comments   Reaction occurs at 338°F and 415 psia using a soluble tellurium-bromine-lithium catalyst.

ETHYLENE OXIDE

Most ethylene oxide is converted to ethylene glycol. Other important derivatives are polyglycols, glycol ethers, ethanolamines, and phenoxyphenols. There are many other derivatives produced on a small scale.

Ethylene oxide was first manufactured from ethylene via ethylene chlorohydrin. In the 1930s a process in which ethylene was oxidized directly to ethylene oxide was introduced. Since then the direct oxidation route has gradually come to the forefront. By 1950 it accounted for nearly one-third the ethylene oxide produced; by 1960 two-thirds was so produced; and today all is derived from this process.

Manufacturing Processes

1 OXIDATION OF ETHYLENE

Material Balance

Chemical	Coefficient T/T Product
Ethylene	−0.96
Ethylene oxide	1.00

Primary energy requirements for utilities	−0.07 FOET/T
Unit investment for a 136 kT plant (1977 $)	0.62 $1000/T

Comments   Reaction occurs in tubular reactors over a silver catalyst in three successive reaction-absorption stages.

2 OXIDATION OF ETHYLENE (OXYGEN)

Material Balance

Chemical	Coefficient T/T Product
Ethylene	−0.88
Ethylene oxide	1.00
Oxygen	−1.10
Miscellaneous chemicals and catalyst	−0.01

Primary energy requirements for utilities	0.01 FOET/T
Unit investment for a 136 kT plant (1977 $)	0.43 $1000/T

Comments   Reaction occurs at about 473°F and 21 atm over a supported silver catalyst packed in tubes of the reactor.

3 CHLOROHYDRATION OF ETHYLENE

Material Balance

Chemical	Coefficient T/T Product
Calcium oxide	−1.47
Chlorine	−1.88
Ethylene	−0.78
Ethylene dichloride	0.18
Ethylene oxide	1.00
Sodium hydroxide	−0.04

Primary energy requirements for utilities	0.47 FOET/T
Unit investment for a 45 kT plant (1977 $)	0.38 $1000/T

2-ETHYLHEXANOL

2-Ethylhexanol is used mainly to produce plastizicers, the most important of which is di-2-ethylhexyl phthalate, used in vinyl chloride resins. Also, substantial quantities of the alcohol are converted to its acrylate ester for use in producing adhesives and surface coatings. It also has some use as a solvent, and has various other small-scale markets.

The aldol condensation of *n*-butyraldehyde is used commercially to produce 2-ethylhexanol. The *n*-butyraldehyde can be derived from acetaldehyde via crotonaldehyde or from propylene by oxonation, the latter being preferred today.

Manufacturing Processes

1 PROPYLENE BY THE OXO PROCESS

Material Balance

Chemical	Coefficient T/T Product
n-Butanol	0.11
Carbon monoxide	−0.63
2-Ethylhexanol	1.00
Fuel oil (low sulfur)	0.45
Hydrogen	−0.08
Isobutanol	0.16
Propylene (chemical grade)	−1.04
Miscellaneous chemicals	−0.25

Primary energy requirements for utilities   0.49 FOET/T
Unit investment for a 64 kT plant (1977 $)   0.56 $1000/T

FORMALDEHYDE

Most formaldehyde is polymerized to form resins such as polyacetal, urea-formaldehyde, phenol-formaldehyde, and melamine-formaldehyde. Other important derivatives are hexamethylenetetramine, used chiefly as a cross-linking agent in phenolic resins, and pentaerythritol, used largely in producing alkyd resins. Substantial quantities of formaldehyde were once used to manufacture ethylene glycol, but this market no longer exists. There are various other formaldehyde derivatives including acrylic acid and tetrahydrofuran.

In the usual route to formaldehyde, methanol is simultaneously oxidized and dehydrogenated over a metal catalyst in such a way that the endothermic dehydrogenation thermally balances the exothermic oxidation. Another process, involving methanol oxidation over a metal oxide catalyst, has long been operated, and now seems to be gaining ground on the thermally balanced process. Some formaldehyde is also produced, along with acetaldehyde and methanol, by oxidizing propane or butane.

Manufacturing Processes

1 OXIDATION OF METHANOL

Material Balance

Chemical	Coefficient T/T Product
Formaldehyde	1.00
Methanol	−1.18

Primary energy requirements for utilities   −0.03 FOET/T
Unit investment for a 45 kT plant (1977 $)   0.27 $1000/T

Comments   Air oxidation of methanol occurs over a molybdenum oxide-iron oxide catalyst at 600–700°F and 5–10 psig.

FUEL GAS

As used here, the term gas refers to light gaseous hydrocarbon fractions which are produced mainly by steam cracking processes used in manufacture of ethylene.

FUEL OIL

As used here, the term fuel oil refers to heavy liquid hydrocarbon fractions (400°F) that can be used to manufacture synthesis gas for the production of ammonia, methanol, and other chemicals. In addition to supplies from refineries, such a liquid fraction is produced as a by-product when gas oil is pyrolyzed to ethylene and other olefins.

GAS OIL

Gas oil is a liquid petroleum fraction (400–600°F) that can serve as a feedstock for the manufacture of synthesis gas and ethylene and other olefins.

GLYCERINE

In its most significant applications glycerine is used as a raw material in the production of alkyd resins, explosives, and food additives, as a humectant and flavoring in tobacco processing, as a softener for cellophane, and as an emulsant for drugs and cosmetics.

Until 1948, when synthetic glycerine was first produced commercially from propylene via allyl chloride and epichlorohydrin, nearly all glycerine was obtained as a by-product of soap manufacture. This is still a major source of glycerine today, accounting for nearly half the glycerine produced. The original synthetic route remains the most important. Other routes involve the oxidation of allyl alcohol with either hydrogen peroxide or peracetic acid. The hydrogen peroxide process has been operated since the late 1950s; the peracetic acid process is a relatively recent development.

Manufacturing Processes

1 ALLYL CHLORIDE VIA EPICHLOROHYDRIN

Material Balance

Chemical	Coefficient T/T Product
Allyl chloride	−1.00
Calcium oxide	−0.78
Chlorine	−0.93
Glycerine	1.00
Hydrogen chloride	−0.08
Sodium carbonate	−0.07
Sodium hydroxide	−0.49
Toluene	−0.01

Primary energy requirements for utilities   0.93 FOET/T
Unit investment for a 45 kT plant (1977 $)   0.55 $1000/T

Comments   Reaction occurs at 320–350°F and 165 psia in the presence of sodium carbonate.

2   HYDROLYSIS OF EPICHLOROHYDRIN

Material Balance

Chemical	Coefficient T/T Product
Epichlorohydrin	−1.03
Glycerine	1.00
Hydrogen chloride	−0.28
Sodium carbonate	−0.07
Sodium hydroxide	−0.49
Toluene	−0.01

Primary energy requirements for utilities   0.64 FOET/T
Unit investment for a 45 kT plant (1977 $)   0.20 $1000/T

3   REACTION OF ALLYL ALCOHOL AND HYDROGEN PEROXIDE

Material Balance

Chemical	Coefficient T/T Product
Allyl alcohol	−0.75
Glycerine	1.00
Hydrogen peroxide	−0.45
Sodium hydroxide	−0.01

Primary energy requirements for utilities   0.38 FOET/T
Unit investment for a 45 kT plant (1977 $)   0.21 $1000/T

HEPTENES

Heptenes, highly branched C_7 olefins are formed by the reaction of isobutylene and propylene in the presence of aluminum chloride or phosphoric acid. They are used mainly in manufacture of C_8 oxo alcohol known as isooctanol.

HEXAMETHYLENEDIAMINE

Hexamethylenediamine is used almost exclusively as a monomer for nylon 66 production. It is manufactured from adiponitrile by hydrogenation.

Manufacturing Processes

1 ELECTROLYTIC DIMERIZATION OF ACRYLONITRILE

Material Balance

Chemical	Coefficient T/T Product
Acrylonitrile	−0.99
Hexamethylenediamine	1.00
Hydrogen	−0.07

Primary energy requirements for utilities   1.65 FOET/T
Unit investment for a 91 kT plant (1977 $)   1.09 $1000/T

2 REACTION OF ADIPIC ACID AND AMMONIA

Material Balance

Chemical	Coefficient T/T Product
Adipic acid	−1.41
Ammonia	−0.49
Hexamethylenediamine	1.00
Hydrogen	−0.07
Sodium bisulfate	−0.12

Primary energy requirements for utilities   0.14 FOET/T
Unit investment for a 32 kT plant (1977 $)   0.81 $1000/T

Comments Reaction occurs at 530°F in presence of a phosphoric acid catalyst.

3 HYDROCYANATION OF BUTADIENE

Material Balance

Chemical	Coefficient T/T Product
Ammonia	−0.06
Butadiene	−0.63
Hexamethylenediamine	1.00
Hydrogen	−0.07
Hydrogen cyanide	−0.61

Primary energy requirements for utilities	0.50 FOET/T
Unit investment for a 91 kT plant (1977 $)	0.80 $1000/T

Comments Butadiene is hydrocyanated to adiponitrile over a soluble nickle-phosphorus catalyst at 100 psia. The resulting adiponitrile is hydrogenated to hexamethylenediamine at 225°F and 5000 psia.

HEXAMETHYLENETETRAMINE

Hexamethylenetetramine was used extensively during World War II and the Korean War, because of its use in the manufacture of the explosive RDX. The petrochemical uses of hexamethylenetetramine are in the manufacture and curing of phenolic resins. Smaller amounts are also used in medicinals, blasting caps, and fuel tablets. Hexamethylenetetramine is manufactured by the reaction of formaldehyde and ammonia.

HYDROGEN

Hydrogen is used in several petrochemical processes, but ammonia and methanol production are by far the largest consumers. For many years hydrogen for methanol and ammonia synthesis was derived from carbon and steam using the water-gas reaction. Today most is produced from hydrocarbons by steam reforming, and, to a lesser extent, by partial oxidation. In either case, methane is the usual feedstock, but heavier hydrocarbons can be used. Carbon monoxide is co-produced in these processes; when the hydrogen is to be used for ammonia synthesis the carbon monoxide is converted to additional hydrogen via the water-gas shift reaction. Hydrogen is also produced in the various dehydrogenation processes used in the petrochemical industry; the quantities so produced may be sufficient to supply allied hydrogenation processes.

Manufacturing Processes

1 STEAM REFORMING OF METHANE

Material Balance

Chemical	Coefficient T/T Product
Hydrogen	1.00
Methane	−2.25

Primary energy requirements for utilities   1.99 FOET/T
Unit investment for a 83 kT plant (1977 $)   0.41 $1000/T

Comments   Reaction occurs at 25 atm and 1600°F.

2 STEAM REFORMING OF NAPHTHA

Material Balance

Chemical	Coefficient T/T Product
Hydrogen	1.00
Naphtha	−2.63

Primary energy requirements for utilities   2.27 FOET/T
Unit investment for a 83 kT plant (1977 $)   0.39 $1000/T

3 PARTIAL OXIDATION OF NAPHTHA

Material Balance

Chemical	Coefficient T/T Product
Carbon dioxide	3.64
Carbon monoxide	7.84
Hydrogen	1.00
Naphtha	−4.17
Oxygen	−4.44
Sulfuric acid	−0.09

Primary energy requirements for utilities   0.56 FOET/T
Unit investment for a 83 kT plant (1977 $)   0.38 $1000/T

HYDROGEN CHLORIDE

Large quantities of hydrogen chloride are produced in petrochemical chlorination processes as a by-product. The surfeit of hydrogen chloride led to the development of oxychlorination processes, which use hydrogen chlorine instead of chlorine as the chlorinating agent. Other than chemical manufacture, the metals industry is the largest consumer of hydrogen chloride.

HYDROGEN CYANIDE

Hydrogen cyanide is consumed mainly in the production of methyl methacrylate and adiponitrile. In the past, large quantities were used to manufacture acrylonitrile, but this market no longer exists, as the cyanide-based processes have been replaced by propylene ammoxidation. Other important derivatives include sodium nitrilotriacetate, used in place of phosphates in some detergent formulations, sodium cyanide, used in electroplating and gold extraction, and cyanuric chloride, used in herbicide manufacture.

Most hydrogen cyanide is produced by the reaction of methane, ammonia, and air. This process was developed in Germany before World War II, but was not used to a significant extent until the early 1950s. Processes that react methane and ammonia but require no air have also been commercialized. One of these processes can be adapted to use propane or naphtha in place of methane. In the traditional route, sodium cyanide produced from ammonia, charcoal, and metallic sodium reacts with sulfuric acid to form hydrogen cyanide and sodium sulfate. Another process once widely used is based on the decomposition of formamide, which is produced from methanol, carbon monoxide, and ammonia. Hydrogen cyanide is produced as a by-product in the manufacture of acrylonitrile from propylene. The amounts so produced are becoming increasingly significant.

Manufacturing Processes

1 AMMOXIDATION OF METHANE

Material Balance

Chemical	Coefficient T/T Product
Ammonia	−0.75
Hydrogen cyanide	1.00
Methane	−1.02
Phosphoric acid	−0.01

Primary energy requirements for utilities   0.13 FOET/T
Unit investment for a 59 kT plant (1977 $)   0.71 $1000/T

Comments Reaction occurs at 30 psia and 2000°F over a platinum-rhodium catalyst.

HYDROGEN PEROXIDE

Aside from its military applications (such as use in rocket fuel), hydrogen peroxide is used primarily for bleaching in the textile and paper industries. Its only large-scale petrochemical use is in one process for producing synthetic glycerine. There are, however, several small-scale chemical derivatives. Hydrogen peroxide can be produced petrochemically from isopropanol and by the hydrogenation and oxidation of anthraquinone. This process is favored today.

Manufacturing Processes

1 ANTHRAQUINONE PROCESS

Material Balance

Chemical	Coefficient T/T Product
Diisobutyl carbinol	−0.02
Hydrogen	−0.06
Hydrogen peroxide	1.00
Methylnaphthalene	−0.02
Miscellaneous	−0.07

Primary energy requirements for utilities   0.77 FOET/T
Unit investment for a 36 kT plant (1977 $)   1.27 $1000/T

Comments Reaction occurs at 25 psia and 95°F over a palladium-on-spinel catalyst.

2 AUTOXIDATION OF ISOPROPANOL

Material Balance

Chemical	Coefficient T/T Product
Acetone	2.11
Hydrogen peroxide	1.00
Isopropanol	−2.26
Miscellaneous	−0.12

Primary energy requirements for utilities   1.43 FOET/T
Unit investment for a 16 kT plant (1977 $)   1.58 $1000/T

Comments   Reaction occurs in liquid phase at 515 psia and 270°F. No catalyst is used.

ISOBUTANE

Isobutane is important primarily in fuel production (it is fed to refinery alkylation units). Significant use of isobutane as a petrochemical feedstock in the United States did not begin until fairly recently. It is cracked to produce isobutylene and propylene, and is converted to isobutylene in one process for making propylene oxide. Dehydrogenation to isobutylene is an old process but was never important in the United States.

Isobutane can be recovered from natural gas or from refinery streams. If these supplies are insufficient, n-butane can be isomerized to isobutane (this process is responsible for much of the isobutane produced in refineries).

Manufacturing Processes

1   ISOMERIZATION OF n-BUTANE

Material Balance

Chemical	Coefficient T/T Product
n-Butane	−1.07
Hydrogen	−0.01
Isobutane	1.00
Pentane	0.02

Primary energy requirements for utilities   0.14 FOET/T
Unit investment for a 124 kT plant (1977 $)   0.05 $1000/T

ISOBUTANOL

There is no single most important use for isobutanol. It and its acetate ester are used as solvents, and it is used to produce lubricating-oil additives, amine resins, plasticizers, and a variety of small-scale chemicals.

In the United Stated today, all isobutanol is derived from the isobutyraldehyde that occurs as a by-product in the oxonation of propylene to n-butyraldehyde. The process for manufacturing ethanol from synthesis gas can be operated to produce by-product isobutanol; this was once the major source of isobutanol. It can also be produced as a by-product in the direct carbonylation of propylene to n-butanol, but this process is not operated in the United States.

ISOBUTYLENE

Currently there are no large-volume petrochemicals derived from isobutylene in the United States. However, methyl methacrylate and isoprene are potential derivatives. Aside from its fuel use (isobutylene is fed to refinery alkylation units), most isobutylene is polymerized to form butyl rubber. Other isobutylene polymers have applications in lubricating oil and in caulking and sealing compounds.

Refinery cracking units are the major suppliers of isobutylene. Some is produced as a by-product in ethylene plants, and it can be derived from isobutane by pyrolysis or catalytic dehydrogenation. In addition isobutane is converted to isobutylene in one process for manufacturing propylene oxide.

Manufacturing Processes

1 ACID EXTRACTION OF STEAM CRACKED BUTENES

Material Balance

Chemical	Coefficient T/T Product
Butenes	−2.53
n-Butylenes	1.51
Isobutylene	1.00

Primary energy requirements for utilities	0.18 FOET/T
Unit investment for a 80 kT plant (1977 $)	0.23 $1000/T

ISOBUTYRALDEHYDE

Isobutyraldehyde is produced as a by-product when propylene is oxonated to n-butyraldehyde. It is usually hydrogenated to isobutanol; this represents virtually the only use for isobutyraldehyde.

ISOOCTANOL

Isooctanol is one of the two principal oxo alcohols. It is produced by oxonation of heptenes in a manner similar to 2-ethylhexanol (the other oxo alcohol) except for the elimination of aldol condensation of intermediate aldehydes. Phthalate plasticizers account for the principal market for both isooctanol and 2-ethylhexanol, but 2-ethyhexanol commands a much larger market than isooctanol.

Manufacturing Processes

1 OXONATION OF HEPTENES (ONE-STEP LOW PRESSURE)

Material Balance

Chemical	Coefficient T/T Product
Acetic acid	−0.01
Carbon monoxide	−0.27
Fuel oil (low sulfur)	0.19
Heptenes	−1.08
Hydrogen	−0.04
Isooctanol	1.00
Sodium hydroxide (dilute)	−0.01
Miscellaneous chemicals and catalyst	−0.02

Primary energy requirements for utilities	0.17 FOET/T
Unit investment for a 64 kT plant (1977 $)	0.30 $1000/T

2 OXONATION OF HEPTENES (TWO-STEP)

Material Balance

Chemical	Coefficient T/T Product
Carbon monoxide	−0.33
Ethylene glycol	−0.01
Fuel oil (low sulfur)	0.41
Heptenes	−1.39
Hydrogen	−0.07
Isooctanol	1.00
Sodium bicarbonate	−0.04
Sulfuric acid	−0.04
Miscellaneous	−0.01

Primary energy requirements for utilities	0.29 FOET/T
Unit investment for a 64 kT plant (1977 $)	0.29 $1000/T

ISOPHTHALIC ACID

Isophthalic acid was first produced commercially in 1956. It is used primarily to form polyesters and alkyd resins. All isophthalic acid is derived from *m*-xylene by oxidation. The original oxidation process involved the intermediate formation of amides or ammonium salts, which were then hydrolyzed to the acid. This has now been supplanted by a direct catalytic oxidation process.

Manufacturing Processes

1 OXIDATION OF *m*-XYLENE

Material Balance

Chemicals	Coefficient T/T Product
Isophthalic acid	1.00
m-Xylene	−0.71

Primary energy requirements for utilities	0.49 FOET/T
Unit investment for a 45 kT plant (1977 $)	0.34 $1000/T

ISOPRENE

Isoprene was not manufactured on a large-scale until the late 1950s, when *cis*-polyisoprene, a synthetic elastomer, was first commercialized. Small quantities of isoprene have been produced previously for copolymerization with isobutylene to form butyl rubber. Today *cis*-polyisoprene is by far the largest market, with butyl rubber a distant second.

In the United States most isoprene is produced from isopentenes or propylene. Elsewhere processes have been operated that derive isoprene from isobutylene and formaldehyde and from acetylene and acetone. Isoprene is also produced as a by-product when naphtha or gas oil is cracked to ethylene. This source is not now important in the United States as most ethylene is derived from ethane or propane. In Europe and Japan, however, where naphtha cracking is the usual route to ethylene, this is a major source of isoprene.

Manufacturing Processes

1 DIMERIZATION OF PROPYLENE

Material Balance

Chemical	Coefficient T/T Product
Fuel oil (low sulfur)	0.64
Isoprene	1.00
Propylene (chemical grade)	−2.08
Miscellaneous	−0.01

Primary energy requirements for utilities	0.99 FOET/T
Unit investment for a 40 kT plant (1977 $)	0.53 $1000/

2 REACTION OF FORMALDEHYDE AND ISOBUTYLENE

Material Balance

Chemical	Coefficient T/T Product
t-Butanol	0.09
Formaldehyde	−0.62
Isobutylene	−0.94
Isoprene	1.00
Sodium hydroxide	−0.01
Sulfuric acid	−0.01
By-product credits	0.13

Primary energy requirements for utilities	0.79 FOET/T
Unit investment for a 36 kT plant (1977 $)	0.42 $1000/T

3 EXTRACTION OF PENTENES

Material Balance

Chemical	Coefficient T/T Product
Isoprene	1.00
Pentenes	−1.00
Miscellaneous chemicals	−0.01

Primary energy requirements for utilities   1.68 FOET/T
Unit investment for a 36 kT plant (1977 $)   0.42 $1000/T

ISOPROPANOL

The largest single use for isopropanol is as an intermediate for acetone production. Glycerine and hydrogen peroxide are other derivatives. It is also an important solvent and is used for the synthesis of a variety of small-volume chemicals.

Isopropanol was the first petrochemical—that is, the first chemical manufactured commercially from a petroleum-derived feedstock. The original process, a two-step hydration of propylene using sulfuric acid, was dominant until fairly recently. Today the direct catalytic hydration of propylene is preferred. This process was first operated in the early 1950s, but its early versions suffered from low conversion per pass.

Manufacturing Processes

1 HYDRATION OF PROPYLENE

Material Balance

Chemical	Coefficient T/T Product
Active carbon	−0.01
Fuel gas	0.02
Isopropanol	1.00
Propylene (chemical grade)	−0.73

Primary energy requirements for utilities   0.39 FOET/T
Unit investment for a 272 kT plant (1977 $)   0.20 $1000/T

Comments Reaction occurs at 490°F and 3000 psia in the presence of a tungsten-containing catalyst.

2 HYDRATION OF PROPYLENE (CATION EXCHANGE RESIN CATALYST)

Material Balance

Chemical	Coefficient T/T Product
Active carbon	−0.01
Isopropanol	1.00
Propylene (chemical grade)	−0.83

Primary energy requirements for utilities   0.38 FOET/T
Unit investment for a 272 kT plant (1977 $)   0.20 $1000/T

MALEIC ANHYDRIDE

The largest market for maleic anhydride is in the production of polyesters. It is also used to produce fumaric acid, alkyd resins, pesticides, plasticizers, and lubricants.

Since its emergence as a large-volume chemical, most maleic anhydride has been produced by oxidizing benzene. From time to time, processes oxidizing n-butylenes to maleic anhydride have also been used. Though this route dates back to the 1930s, there is continued interest in it today. There is also current interest in using a butane feedstock.

Manufacturing Processes

1 OXIDATION OF BENZENE

Material Balance

Chemical	Coefficient T/T Product
Benzene	−1.19
Maleic anhydride	1.00

Primary energy requirements for utilities   0.15 FOET/T
Unit investment for a 27 kT plant (1977 $)   0.91 $1000/T

Comments   Reaction occurs at 750° over a vanadium pentoxide-based catalyst in a fixed bed reactor.

2 OXIDATION OF n-BUTANE

Material Balance

Chemical	Coefficient T/T Product
n-Butane	−1.22
Maleic anhydride	1.00

Primary energy requirements for utilities   0.41 FOET/T
Unit investment for a 27 kT plant (1977 $)   1.14 $1000/T

Comments   An oxidation of *n*-butane occurs over a phosphorus-vanadium-zinc catalyst in a fixed-bed reactor.

MELAMINE

Most of the melamine produced is condensed with formaldehyde or other aldehydes, forming resins with excellent resistance to heat, water, and most chemicals, plus good electrical properties and surface hardness. The major outlet for melamine is in high pressure laminates for use in home furnishings, such as decorative countertops and paneling. Melamine resins are also used in areas such as dinnerware, paper and textile treating, adhesives, and surface coatings.

Production of melamine in the United States is exclusively via the urea process, with most derived by the one-step Stamicarbon (DSM) process.

Manufacturing Processes

1   BASF PROCESS

Material Balance

Chemical	Coefficient T/T Product
Ammonia	0.86
Melamine	1.00
Urea	−3.04
Miscellaneous chemicals and catalyst	−0.01

Primary energy requirements for utilities	0.69 FOET/T
Unit investment for a 40 kT plant (1977 $)	0.85 $1000/T

2   CHEMIE LINZ PROCESS

Material Balance

Chemical	Coefficient T/T Product
Ammonia	0.88
Melamine	1.00
Urea	−3.10

Primary energy requirements for utilities	0.87 FOET/T
Unit investment for a 40 kT plant (1977 $)	0.90 $1000/T

3 NISSAN PROCESS

Material Balance

Chemical	Coefficient T/T Product
Ammonia	0.85
Melamine	1.00
Urea	−3.13

Primary energy requirements for utilities   0.71 FOET/T
Unit investment for a 40 kT plant (1977 $)   0.99 $1000/T

4 STAMICARBON PROCESS

Material Balance

Chemical	Coefficient T/T Product
Ammonia	0.84
Melamine	1.00
Urea	−3.04
Miscellaneous chemicals and catalyst	−0.01

Primary energy requirements for utilities   0.93 FOET/T
Unit investment for a 40 kT plant (1977 $)   0.89 $1000/T

METHANE

Aside from its use as fuel, methane is important primarly as a feedstock for synthesis gas production. It is also used to produce acetylene, carbon disulfide, hydrogen cyanide, and the chloromethanes. Natural gas is by far the most important source of methane. Substantial quantities are produced as a by-product of ethylene manufacture but this is not usually used as a feedstock.

METHANOL

Methanol has some market as a solvent but is used primarily as a chemical intermediate. Formaldehyde production consumes the largest amount of methanol, substantial amounts are used to manufacture dimethyl terephthalate, acetic acid, methyl methacrylate, methyl amines, and methyl halides. Other outlets include its use as antifreeze and as fuel.

Methanol is produced by hydrogenating carbon monoxide, carbon dioxide, or mixtures thereof. The preferred feedstock is a synthesis gas. Traditionally this process has been operated at quite high pressure (275-360 atm), but recently processes using lower pressure (50-60 atm) have been gaining favor. Some methanol is still produced by the destructive distillation of wood, but compared to the amounts produced synthetically this source has not been significant for decades.

Manufacturing Processes

1   FROM METHANE (INTERMEDIATE PRESSURE)

Material Balance

Chemical	Coefficient T/T Product
Methane	−0.49
Methanol	1.00

Primary energy requirements for utilities   0.40 FOET/T
Unit investment for a 318 kT plant (1977 $)   0.17 $1000/T

Comments   Reaction occurs at 103 atm and 450–520°F using a copper oxide, aluminum oxide, and zinc oxide catalyst.

2   HYDROGENATION OF CARBON MONOXIDE (HIGH PRESSURE)

Material Balance

Chemical	Coefficient T/T Product
Carbon dioxide	−0.32
Methanol	1.00
Synthesis gas (H_2:CO = 3:1)	−0.92

Primary energy requirements for utilities   0.23 FOET/T
Unit investment for a 318 kT plant (1977 $)   0.20 $1000/T

3   HYDROGENATION OF CARBON MONOXIDE (LOW PRESSURE)

Material Balance

Chemical	Coefficient T/T Product
Carbon dioxide	−0.36
Methanol	1.00
Sulfuric acid	−0.02
Synthesis gas (H_2:CO = 3:1)	−0.89

Primary energy requirements for utilities   0.25 FOET/T
Unit investment for a 318 kT plant (1977 $)   0.19 $1000/T

METHYL ACETATE

Methyl acetate is used mainly as a paint remover, but being flammable, it is less desirable than methylene chloride. Methyl acetate is obtained by the polyvinyl alcohol producers as a mixture with methanol. It can also be produced as a by-product of acetic acid production from acetaldehyde.

METHYL ACRYLATE

Methyl acrylate is manufactured from acrylic acid and consumed in production of polyacrylates. It is also used with acrylonitrile in the manufacture of some acrylic fibers to improve dyeability.

Manufacturing Processes

1   ESTERIFICATION OF ACRYLIC ACID

Material Balance

Chemical	Coefficient T/T Product
Acrylic acid	−0.88
Methanol	−0.39
Methyl acrylate	1.00
Sodium hydroxide	−0.07
Sulfuric acid	−0.06

Primary energy requirements for utilities   0.39 FOET/T
Unit investment for a 45 kT plant (1977 $)   0.30 $1000/T

METHYL CHLORIDE

The two largest markets for methyl chloride are the production of tetramethyl lead for gasoline antiknock fluid and the production of methylchlorosilanes for silicone resins. Substantial quantities are also consumed in manufacturing the other chloromethanes, especially methylene dichloride and chloroform. It is also used as a solvent in butyl rubber manufacture, and at one time was important as a refrigerant.

Most methyl chloride is produced by the hydrochlorination of methanol. Methane can be chlorinated to methyl chloride; this may be favored if the manufacturer has an outlet for the hydrogen chloride that is co-produced and for the other chloromethanes that can be co-produced.

METHYLENE CHLORIDE

Methylene chloride is used primarily as a paint remover. Substantial quantities are also used in solvent degreasing and plastics processing. As a gas it is used as a propellant in aerosol formulations. There is also a sizable export market.

Methane or methanol-derived methyl chloride can be chlorinated to methylene chloride. Both routes have been important, but today methane chlorination is dominant.

METHYLENE DIPHENYLENE DIISOCYANATE (MDI) AND ITS POLYMERIC FORM (PMPPI)

The continually expanding acceptance of polyurethane products in a diverse range of applications has resulted in isocyanates becoming one of the more rapidly growing major petrochemical intermediates. The major isocyanates used for polyurethanes are toluene diisocyanate and polymeric isocyanates, which constitute about 65% and 30% of the United States' requirements, respectively.

Presently, the polyisocyanates are produced by the phosgenation of polyamines, which in turn, are produced by the condensation of aniline and formaldehyde. The newest route to polymeric isocyanates has been developed by Arco Chemicals, but no commercial plant currently exists. This process eliminates polyamines completely, and is based on selenium catalyzed direct carbonylation of nitrobenzene.

Manufacturing Processes

1 REACTION OF ANILINE AND FORMALDEHYDE

Material Balance

Chemical	Coefficient T/T Product
Aniline	−0.80
Chlorobenzene	−0.03
Formaldehyde	−0.16
Hydrogen chloride	0.46
Phosgene	−0.84
Polyisocyanates (MDI/PMPPI)	1.00
Sodium hydroxide	−0.18

Primary energy requirements for utilities   0.36 FOET/T
Unit investment for a 45 kT plant (1977 $)   0.63 $1000/T

METHYL ETHYL KETONE

Methyl ethyl ketone is used largely as a solvent for vinyl coatings, nitrocellulose, adhesives, and acrylic coatings. A variety of other solvent uses accounts for the remainder of the market.

Most methyl ethyl ketone is still produced by dehydrogenating s-butanol. However, the oxidation of butane today accounts for an increasing fraction of total methyl ethyl ketone production. The conversion of n-butylenes to methyl ethyl ketone using the Wacker process for olefin oxidation has also been practiced recently.

Manufacturing Processes

1 DEHYDROGENATION OF s-BUTANOL

Material Balance

Chemical	Coefficient T/T Product
s-butanol	−1.10
Methyl ethyl ketone	1.00
Nitrogen	−0.09

Primary energy requirements for utilities   1.00 FOET/T
Unit investment for a 45 kT plant (1977 $)   0.15 $1000/T

2 OXIDATION OF n-BUTYLENES

Material Balance

Chemical	Coefficient T/T Product
n-Butylenes	−0.84
Methyl ethyl ketone	1.00
sec-Butyl ether	0.12

Primary energy requirements for utilities   0.48 FOET/T
Unit investment for a 45 kT plant (1977 $)   0.52 $1000/T

METHYL ISOBUTYL KETONE

Most methyl isobutyl ketone is used as a solvent, primarily for nitrocellulose, vinyl coatings, and various other coatings and films. Methyl isobutyl ketone is derived from acetone via diacetone alcohol and mesityl oxide. Processes proceeding from acetone or isopropanol directly to methyl isobutyl ketone have recently been developed but apparently not yet commercialized.

Manufacturing Processes

1 DIRECT CONDENSATION OF ACETONE

Material Balance

Chemical	Coefficient T/T Product
Acetone	−1.22
Methyl isobutyl ketone	1.00

Primary energy requirements for utilities   0.25 FOET/T
Unit investment for a 23 kT plant (1977 $)   0.21 $1000/T

METHYL METHACRYLATE

The largest market for methyl methacrylate is in the production of polymers such as Lucite and Plexiglas, which because of their clarity and strength are used as alternatives to glass in various situations. Methyl methacrylate polymers and copoymers are also used extensively as surface coatings, as oil additives, and as molding powders for automobile parts.

Methyl methacrylate is derived from acetone via acetone cyanohydrin. This was the original commercial route to methyl methacrylate and is the only route used today in the United States. An alternate process, developed in the late 1950s and used here in the 1960s, converts isobutylene to methyl methacrylate.

Manufacturing Processes

1 ACETONE CYANOHYDRIN PROCESS

Material Balance

Chemical	Coefficient T/T Product
Acetone	−0.68
Hydrogen cyanide	−0.32
Methanol	−0.37
Methyl methacrylate	1.00
Sulfuric acid	−1.63

Primary energy requirements for utilities 0.24 FOET/T
Unit investment for a 45 kT plant (1977 $) 0.82 $1000/T

2 FROM ISOBUTYLENE VIA METHACRYLIC ACID

Material Balance

Chemical	Coefficient T/T Product
Isobutylene	−1.12
Methanol	−0.38
Methyl methacrylate	1.00
Pentane	−0.03
Sulfuric acid	−0.01
Miscellaneous chemicals and catalyst	−0.01

Primary energy requirements for utilities 0.40 FOET/T
Unit investment for a 45 kT plant (1977 $) 0.85 $1000/T

Comments Oxidation of isobutylene occurs at 788°F and 23 psia in the presence of a molybdenum-cobalt-tungsten catalyst. The resulting methacrolein from the first step is further oxidized to methacrylic acid at 644°F and 23 psia in the presence of a molybdenum-phosphorus catalyst. The purified methacrylic acid is esterified with methanol in the presence of sulfuric acid.

NAPHTHA

Naphtha is a liquid petroleum fraction (C_5–400 F) that can serve as a feedstock for the production of ethylene, acetylene, synthesis gas, and other chemicals. It has been important as feedstock for petrochemicals in Europe and Japan for several years, and is increasingly important in the United States.

NAPHTHALENE

Production of phthalic anhydride has long been the largest market for naphthalene. Among smaller-volume naphthalene derivatives are insecticides, dyestuffs, and tanning agents. Until the early 1960s coal carbonization processes accounted for the entire naphthalene supply. Since then the dealkylation of refinery reformate streams containing methyl naphthalenes has become an equally important source.

NITRIC ACID

Fertilizer production is by far the largest market for nitric acid. Its other important derivatives are explosives such as TNT and petrochemicals such as aniline and toluene diisocyanate. Nearly all nitric acid is manufactured from ammonia by oxidation.

Manufacturing Processes

1 OXIDATION OF AMMONIA (UHDE-TYPE PROCESS)

Material Balance

Chemical	Coefficient T/T Product
Ammonia	−0.28
Nitric acid, 95%	1.00
Oxygen	−0.17

Primary energy requirements for utilities	0.06 FOET/T
Unit investment for a 66 kT plant (1977 $)	0.18 $1000/T

2 OXIDATION OF AMMONIA (GRANDE PAROISSE-TYPE PROCESS)

Material Balance

Chemical	Coefficient T/T Product
Ammonia	−0.28
Nitric acid, 60%	1.00

Primary energy requirements for utilities	0.00 FOET/T
Unit investment for a 181 kT plant (1977 $)	0.06 $1000/T

Comments   Reaction occurs at about 1380°F and 7.8 atm over a platinum-rhodium catalyst.

NITROBENZENE

Most nitrobenzene is hydrogenated to aniline. Other dye intermediates, such as benzidene, are also derived from nitrobenzene. There are small-volume uses for nitrobenzene as an intermediate in the production of pharmaceuticals, herbicides, and polyamide fibers. All nitrobenzene is manufactured by nitrating benzene with an aqueous solution of nitric and sulfuric acids.

Manufacturing Processes

1 NITRATION OF BENZENE

Material Balance

Chemical	Coefficient T/T Product
Benzene	−0.66
Nitric acid, 60%	−0.54
Nitrobenzene	1.00
Sulfuric acid	−0.03

Primary energy requirements for utilities	0.15 FOET/T
Unit investment for a 68 kT plant (1977 $)	0.21 $1000/T

NITROGEN

Nitrogen is one of the fastest growing chemicals produced in the United States. One of the major uses of nitrogen is for blanketing purposes for exclusion of oxygen and moisture. Its other main use is to obtain extremely low temperatures, down to -345°F.

OLEUM

Oleums are strong acids resulting from dissolving SO_3 in sulfuric acid (H_2SO_4). Substantial quantities of oleums are used in production of caprolactam and in many other chemical operations such as phenol, nitrocellulose, nitroglycerine, TNT and dye manufacture.

Commercial oleums fall into three categories: 10-35% free SO_3, 40% free SO_3, and 60-65% SO_3. The 20% to 35% grades can readily be produced in a single absorption tower, whereas 40% oleum generally requires two absorption towers in series. Oleum containing 60-65% free SO_3 must be made by distilling

SO_3 gas out of 20–35% oleum, condensing the 100% SO_3, and then mixing it with additional 20–35% oleum.

OXYGEN

The major market for oxygen is in steel manufacturing followed by its use in cutting and welding of metals. Oxygen is being considered for any chemical process that requires air as an oxidant. It is simply a matter of economics to decide at what point the use of the more efficient oxygen becomes more economical than that of free but less efficient air. In addition, uses of oxygen for wastewater treatment, and for pulp and paper making have opened up entirely new markets for this substance.

Commercial oxygen is produced by the liquefaction and subsequent fractionation of air in several cycles depending on the methods of air compression, purification, and refrigeration.

PENTENES

Pentenes, obtained from refinery cracking units, are used in the petrochemical industry to produce isoprene. This is the only large-scale use for pentenes as a chemical feedstock.

PHENOL

Phenolic resins, especially phenol-formaldehyde, account for about half the demand for phenol. There is also a substantial market for phenol as an intermediate for the production of bisphenol-A and cyclohexanol. Among other phenol derivatives are a variety of surfactants, antioxidants, pesticides, and plasticizers.

The oldest commercial routes to phenol involve the sulfonation of benzene and the alkaline hydrolysis of chlorobenzene. In 1940 a process in which chlorobenzene is catalytically hydrolyzed with steam was introduced, but did not exhibit substantial growth until the 1950s. The hydrochloric acid coproduced in this process is recycled to chlorobenzene production. A process oxidizing cumene to phenol and acetone was commercialized in 1955, and soon assumed a dominant position. Since the early 1960s some phenol has also been derived from toluene via benzoic acid. From time to time, processes oxidizing benzene directly to phenol have been operated, but these did not prove successful. There has also been some interest in the commercial production of phenol from cyclohexanol.

Manufacturing Processes

1 OXIDATION OF CUMENE

Material Balance

Chemical	Coefficient T/T Product
Acetone	0.61
Cumene	−1.35
Phenol	1.00
Sodium hydroxide	−0.01
Sulfuric acid	−0.01
Miscellaneous chemicals	−0.01

Primary energy requirements for utilities	0.38 FOET/T
Unit investment for a 91 kT plant (1977 $)	0.49 $1000/T

Comments Air oxidation of cumene occurs at 217–235°F and 80 psig.

2 DEHYDROCHLORINATION OF CHLOROBENZENE

Material Balance

Chemical	Coefficient T/T Product
Chlorobenzene	−1.19
Hydrogen chloride	0.48
Phenol	1.00

Primary energy requirements for utilities	0.53 FOET/T
Unit investment for a 45 kT plant (1977 $)	0.55 $1000/T

3 ALKALINE HYDROLYSIS OF CHLOROBENZENE

Material Balance

Chemical	Coefficient T/T Product
Chlorobenzene	−1.31
Hydrogen chloride (dilute)	−0.47
Phenol	1.00
Sodium hydroxide	−1.06

Primary energy requirements for utilities   0.48 FOET/T
Unit investment for a 45 kT plant (1977 $)   0.48 $1000/T

4 SULFONATION OF BENZENE

Material Balance

Chemical	Coefficient T/T Product
Benzene	−0.94
Oleum	−1.19
Phenol	1.00
Sodium hydroxide	−1.64
Sodium sulfite	1.93

Primary energy requirements for utilities   0.21 FOET/T
Unit investment for a 32 kT plant (1977 $)   0.58 $1000/T

PHOSGENE

Phosgene had no large-scale commercial use until the mid-1950s, when the manufacture of isocyanates for polyurethane production began. Isocyanates still account for most of the demand for phosgene. Other important phosgene derivatives are polycarbonate resins and the insecticide Sevin. The only commercial route to phosgene is the reaction of chlorine and carbon monoxide.

Manufacturing Processes

1 REACTION OF CHLORINE AND CARBON MONOXIDE

Material Balance

Chemical	Coefficient T/T Product
Carbon monoxide	−0.30
Chlorine	−0.73
Phosgene	1.00

Primary energy requirements for utilities   0.07 FOET/T
Unit investment for a 61 kT plant (1977 $)   0.20 $1000/T

Comments   Reaction occurs over an active carbon catalyst.

PHOSPHORIC ACID

Phosphoric acid is a major inorganic acid; close to 85% is used to manufacture superphosphates, ammonium phosphates, and dicalcium phosphate for fertilizers. Most of the remaining material is used in the manufacture of alkali phosphates for cleaning compounds and detergents. A fairly large direct use of phosphoric acid is in soft cola drinks, where it adds a characteristic tart flavor. Other direct uses include rust-proofing, electropolishing of stainless steel and aluminum, petroleum refining, and catalysts for olefin polymerization and alkylation.

The oldest method of manufacturing phosphoric acid is the wet process, which consists of treating phosphate rock with sulfuric acid and releasing the free phosphoric acid. The newer, furnace method involves burning elemental phosphorus to phosphorus pentoxide, which is then hydrated to phosphoric acid.

PHTHALIC ANHYDRIDE

About half the total production of phthalic anhydride is used to manufacture plasticizers. Other major markets are the production of polyesters and alkyd resins. Phthalic anhydride is produced commercially by oxidizing either naphthalene or o-xylene. Although some phthalic anhydride has been derived from o-xylene since 1945, it was not until substantial quantities of o-xylene became available in the late 1950s that the process became important. o-xylene is now the preferred feedstock, but it did not overtake naphthalene until the early 1970s.

Manufacturing Processes

1 OXIDATION OF o-XYLENE

Material Balance

Chemical	Coefficient T/T Product
Phthalic anhydride	1.00
o-Xylene	−0.97

Primary energy requirements for utilities	0.07 FOET/T
Unit investment for a 32 kT plant (1977 $)	0.58 $1000/T

Comments An oxidation of o-xylene occurs at about 716°F and 27 psia over a vanadium pentoxide-titanium dioxide catalyst contained in fixed-bed reactors.

2 OXIDATION OF NAPHTHALENE

Material Balance

Chemical	Coefficient T/T Product
Naphthalene	−1.02
Phthalic anhydride	1.00

Primary energy requirements for utilities	−0.12 FOET/T
Unit investment for a 32 kT plant (1977 $)	0.38 $1000/T

PROPANE

Aside from its use as fuel, propane is important primarily as a feedstock for ethylene manufacture, though it can also be used to produce acetylene, synthesis gas, perchloroethylene, and other chemicals. Natural gas is the source of most of the propane consumed by the petrochemical industry; refinery gas streams account for the remainder.

PROPYLENE

Aside from fuel production (propylene is fed to refinery alkylation and polymerization units), most propylene is used as a chemical intermediate. Isopropanol, acrylonitrile, cumene, propylene oxide, n-butyraldehyde, and allyl chloride are its major derivatives. From these is produced a wide range of other chemicals, including acetone, phenol, n-butanol, 2-ethylhexanol, glycerine, epichlorohydrin, propylene glycol, and adiponitrile. Large quantities of propylene are polymerized to polypropylene, a light-weight plastic, and to propylene trimer and tetramer, feedstocks for synthetic detergent manufacture. Propylene is also polymerized with ethylene to yield elastomers.

Refinery cracking units are the major suppliers of propylene. Large quantities are also recovered as a by-product of ethylene manufacture.

Manufacturing Processes

1 CHEMICAL GRADE PROPYLENE FROM REFINERY GRADE PROPYLENE

Material Balance

Chemical	Coefficient T/T Product
Propane	0.33
Propylene (chemical grade)	1.00
Propylene (refinery grade)	−1.33

Primary energy requirements for utilities   0.01 FOET/T
Unit investment for a 181 kT plant (1977 $)   0.03 $1000/T

2 POLYMER GRADE PROPYLENE FROM REFINERY GRADE PROPYLENE

Material Balance

Chemical	Coefficient T/T Product
Propane	0.43
Propylene (polymer grade)	1.00
Propylene (refinery grade)	−1.43

Primary energy requirements for utilities   0.01 FOET/T
Unit investment for a 181 kT plant (1977 $)   0.04 $1000/T

PROPYLENE DICHLORIDE

Propylene dichloride is produced as a by-product in the chlorohydrin route to propylene oxide. It has a small market as a solvent and as a lead scavenger in leaded gasoline. Because the amount produced far exceeds this demand, it is usually disposed by converting it to perchloroethylene and carbon tetrachloride by chlorinolysis. Should the need arise, propylene dichloride could also be produced by chlorinating propylene.

PROPYLENE GLYCOL

The largest market for propylene glycol is the production of polyester resins. Other important outlets are cellophane production, tobacco processing, and plasticizer production. Propylene glycol is produced commercially by hydrating propylene oxide. New technologies in which propylene glycol is derived directly from propylene are being developed.

Manufacturing Processes

1 HYDRATION OF PROPYLENE OXIDE

Material Balance

Chemical	Coefficient T/T Product
Dipropylene glycol	0.10
Propylene glycol	1.00
Propylene oxide	−0.89

Primary energy requirements for utilities   0.26 FOET/T
Unit investment for a 45 kT plant (1977 $)   0.20 $1000/T

PROPYLENE OXIDE

The two major markets for propylene oxide are the production of propylene
glycol for polyester resins and the production of polypropoxyethers or polyols
for polyurethane foams. It also has various small-scale uses as a solvent and as a
chemical intermediate.

   Until recently all propylene oxide manufactured commercially was produced
from propylene via propylene chlorohydrin in a process similar to that once used
to make ethylene oxide. In a process commercialized in the late 1960s,
isobutane is oxidized to t-butyl hydroperoxide, which is then used to oxidize
propylene to propylene oxide. In the process, the t-butyl hydroperoxide is converted to t-butanol. Because there is little use for the latter, it is routinely
dehydrated to isobutylene. Ethylbenzene has been used in place of isobutane; in
this case styrene replaced isobutylene as the co-product.

Manufacturing Processes

1 CHLOROHYDRATION OF PROPYLENE

Material Balance

Chemical	Coefficient T/T Product
Ammonia	−0.01
Calcium oxide	−1.20
Chlorine	−1.46
Hydrogen chloride	−0.03
Propylene (chemical grade)	−0.83
Propylene dichloride	0.10
Propylene oxide	1.00
Miscellaneous chemicals	−0.01

Primary energy requirements for utilities   0.41 FOET/T
Unit investment for a 181 kT plant (1977 $)   0.34 $1000/T

2 OXIDATION OF PROPYLENE WITH *t*-BUTYLHYDROPEROXIDE

Material Balance

Chemical	Coefficient T/T Product
Fuel oil (low sulfur)	0.23
Isobutane	−2.61
Isobutylene	2.15
Oxygen	−1.13
Propylene (chemical grade)	−0.78
Propylene oxide	1.00

Primary energy requirements for utilities	0.83 FOET/T
Unit investment for a 181 kT plant (1977 $)	0.87 $1000/T

PYROLYSIS GASOLINE

A light hydrocarbon liquid (C_5–400 F), known as pyrolysis gasoline, is produced as a by-product when hydrocarbons are pyrolyzed to ethylene. This liquid is rich in aromatics and in isoprene. Large quantities of pyrolysis gasoline are produced when naphtha or gas oil is used for ethylene manufacture. Marketing this by-product may be difficult unless the ethylene manufacturer is allied with a petroleum refinery where the pyrolysis gasoline can be blended in motor gasoline or used in other operations.

SODIUM BICARBONATE

Sodium bicarbonate is produced by treating a saturated solution of sodium carbonate with carbon dioxide. Production of this chemical has been fairly stable for many years. Its market strength is primarily in food and pharmaceutical areas, but it is also consumed in areas such as fire extinguishers, and soap and detergents.

SODIUM BISULFATE

Sodium bisulfate is an easily handled dry material which reacts as sulfuric acid. Its major uses are in the manufacture of acid-type toilet bowl cleaners and for industrial cleaning and metal pickling. It is also used in dye baths, carbonizing wool, and in production of various chemicals such as adiponitrile and hexamethylenediamine.

Sodium bisulfate is commonly called "niter cake" because it was obtained by the obsolete process of reacting sodium nitrate with sulfuric acid. It may also be produced by heating sodium chloride (salt) with sulfuric acid.

SODIUM CARBONATE

Sodium carbonate is most used for its properties as a strong alkali, and so is in direct competition with sodium hydroxide for many uses. Its market in the aluminum industry has already been lost to sodium hydroxide, but a tight supply situation in sodium hydroxide has enabled sodium carbonate to maintain its market in the glass industry.

The ammonia-soda or Solvay process for production of sodium carbonate was dominant for many years, but no Solvay plants have been built since 1934. The usual source of sodium carbonate, however, is Trona deposits (natural sesquicarbonate) or lake brines.

SODIUM CHLORIDE

Sodium chloride is produced in processes currently used to manufacture chloroprene and glycerine, as well as in obsolete processes for phenol and vinyl chloride. The sodium chloride so produced is usually electrolyzed to recover the chlorine. The usual source of sodium chloride is sea water or other natural brines.

SODIUM HYDROXIDE

Sodium hydroxide is consumed in current processes for manufacturing chloroprene and glycerine and in obsolete processes for phenol and vinyl chloride. This represents a very small fraction of the total demand for this chemical. The electrolysis of sodium chloride accounts for most of the sodium hydroxide supply.

SODIUM SULFITE

Sodium sulfite is produced in a process for making phenol. It is usually produced by the reaction of sulfur dioxide and sodium carbonate. It is used mainly for bleaching and as a preservative for foods and beverages.

SORBITOL

The largest market for sorbitol is in toothpaste and toiletries manufacture, where it serves as a humectant and a gel base. Ascorbic acid (vitamin C) is the second largest market for this material. Consumption of sorbitol in the petrochemical industry is limited to manufacture of polyether polyols, which in turn, are used in production of polyurethane foams.

The first commercial plant for the production of sorbitol was built in 1937, based on an electrolytic reduction process using a dextrose (also called glucose or corn sugar) solution as feedstock. In 1947 the electrolytic plant was shut down and replaced by a catalytic hydrogenation plant. All expansions since then have been of a catalytic nature.

STYRENE

Production of polystyrene accounts for about half the market for styrene. Most of the remainder is used to produce copolymers, of which the most important are styrene-butadiene, acrylonitrile-butadiene-styrene, and styrene-acrylonitrile. There is also a substantial export market.

Nearly all styrene is manufactured from ethylbenzene by dehydrogenation. An alternative process oxidizes ethylbenzene to acetophenone, which is then reduced to phenyl ethyl alcohol for dehydration to styrene. This process has been operated in the United States, but now is used only in Europe. Another source of styrene is the pyrolysis gasoline produced in ethylene manufacture. Also styrene is co-produced in one process to manufacture propylene oxide.

Manufacturing Processes

1 DEHYDROGENATION OF ETHYLBENZENE

Material Balance

Chemical	Coefficient T/T Product
Benzene	0.03
Ethylbenzene	−1.15
Styrene	1.00
Toluene	0.05

Primary energy requirements for utilities	0.32 FOET/T
Unit investment for a 454 kT plant (1977 $)	0.19 $1000/T

Comments   Reaction occurs at about 1100°F and 10 psig in the presence of an iron-chromium-potassium catalyst.

2 ETHYLBENZENE BY HYDROPEROXIDE PROCESS

Material Balance

Chemical	Coefficient T/T Product
Ethylbenzene	−1.14
Propylene (chemical grade)	−0.33
Propylene oxide	0.41
Sodium hydroxide	−0.01
Styrene	1.00

Primary energy requirements for utilities	0.43 FOET/T
Unit investment for a 454 kT plant (1977 $)	0.37 $1000/T

Comments   Air oxidation of ethylbenzene occurs in liquid phase at 284°F. The ethylbenzene hydroperoxide solution reacts with propylene in the presence of a molybdenum catalyst at 230°F and 600 psig.

SULFUR

Elemental sulfur is consumed by the petrochemical industry in producing carbon disulfide and in preparing hydroxylamine sulfate for caprolactam production. These are its only large-scale petrochemical applications as a raw material. Elemental sulfur is obtained from natural deposits or from naturally occurring hydrogen sulfide gas.

SULFURIC ACID

Sulfuric acid is consumed in current processes for manufacturing caprolactam and methyl methacrylate, and is used in an auxiliary capacity in many petrochemical operations. These applications account for a very small fraction of the total demand for sulfuric acid. Most sulfuric acid is produced by the stepwise oxidation of sulfur to sulfur trioxide, which is then contacted with water to yield sulfuric acid.

Manufacturing Processes

1 DOUBLE ABSORPTION PROCESS

Material Balance

| | Coefficient |
Chemical	T/T Product
Sulfur	-0.33
Sulfuric acid	1.00

Primary energy requirements for utilities   0.01 FOET/T
Unit investment for a 306 kT plant (1977 $)   0.05 $1000/T

SYNTHESIS GAS

Synthesis gas, a mixture of hydrogen and carbon monoxide, is commonly used in the production of ammonia and methanol. Synthesis gas can be converted to liquid hydrocarbons by the Fischer-Tropsch process. This process dates back to 1923, and comprises carbon monoxide and hydrogen condensation at 325–425°C and 100–150 atm over K_2CO_3 coated iron turnings. A mixture of alcohols, acids, esters, and so on, are produced under these conditions.

Manufacturing Processes

1 METHANE REFORMING

Material Balance

| | Coefficient |
Chemical	T/T Product
Methane	-0.55
Synthesis gas (H_2:CO = 1:1)	1.00

Primary energy requirements for utilities   0.12 FOET/T
Unit investment for a 1265 kT plant (1977 $)   0.18 $1000/T

2 PARTIAL OXIDATION OF RESIDUAL OIL

Material Balance

Chemical	Coefficient T/T Product
Fuel oil (high sulfur)	−0.52
Oxygen	−0.53
Synthesis gas (H_2:CO = 1:1)	1.00

Primary energy requirements for utilities	−0.06 FOET/T
Unit investment for a 1265 kT plant (1977 $)	0.06 $1000/T

3 COAL GASIFICATION (LURGI PROCESS)

Material Balance

Chemical	Coefficient T/T Product
Ammonia	0.02
Coal	−2.11
Sulfur	0.09
Synthesis gas (H_2:CO = 2:1)	1.00

Primary energy requirements for utilities	−0.15 FOET/T
Unit investment for a 900 kT plant (1977 $)	0.28 $1000/T

4 STEAM REFORMING OF NAPHTHA

Material Balance

Chemical	Coefficient T/T Product
Naphtha	−0.41
Synthesis gas (H_2:CO = 2:1)	1.00

Primary energy requirements for utilities	0.39 FOET/T
Unit investment for a 900 kT plant (1977 $)	0.15 $1000/T

5 PARTIAL OXIDATION OF RESIDUAL OIL

Material Balance

Chemical	Coefficient T/T Product
Fuel oil (high sulfur)	−0.73
Oxygen	−0.74
Synthesis gas ($H_2:CO = 2:1$)	1.00

Primary energy requirements for utilities	0.03 FOET/T
Unit investment for a 900 kT plant (1977 $)	0.09 $1000/T

6 COAL GASIFICATION (LURGI PROCESS)

Material Balance

Chemical	Coefficient T/T Product
Ammonia	0.03
Coal	−2.48
Sulfur	0.11
Synthesis gas ($H_2:CO = 3:1$)	1.00

Primary energy requirements for utilities	−0.17 FOET/T
Unit investment for a 720 kT plant (1977 $)	0.32 $1000/T

7 PARTIAL OXIDATION OF RESIDUAL OIL

Material Balance

Chemical	Coefficient T/T Product
Fuel oil (high sulfur)	−0.91
Oxygen	−0.93
Synthesis gas ($H_2:CO = 3:1$)	1.00

Primary energy requirements for utilities	0.10 FOET/T
Unit investment for a 720 kT plant (1977 $)	0.12 $1000/T

8 METHANE REFORMING

Material Balance

Chemical	Coefficient T/T Product
Methane	−0.49
Synthesis gas (H$_2$:CO = 3:1)	1.00

Primary energy requirements for utilities	0.38 FOET/T
Unit investment for a 720 kT plant (1977 $)	0.12 $1000/T

TEREPHTHALIC ACID

Terephthalic acid (TPA) and its dimethyl ester (DMT) are used almost entirely for the production of polyethylene terephthalate, a polyester first prepared commercially in the United States in the early 1950s. Because TPA was less readily purified than DMT, it was not used directly for polyester manufacture until the mid-1960s, when purification processes were sufficiently improved. Since then the fraction of polyethylene terephthalate manufactured directly from TPA has gradually increased.

In the United States all TPA is produced by oxidizing *p*-xylene with either nitric acid or oxygen. Nitric acid was originally used, but the oxygen-based process has been predominant since the mid-1960s. There are three variants of the oxygen-based process; they differ primarily in whether bromine, methyl ethyl ketone, or acetaldehyde is used as a catalyst promoter. Some of the methyl ethyl ketone or acetaldehyde used as a promoter is converted to acetic acid in the process. TPA can also be derived from benzoic acid obtained from toluene or phthalic anhydride. This process is important in Japan, where there are insufficient supplies of *p*-xylene. A recently developed process involves the ammonolysis of *p*-xylene to terephthalonitrile, followed by hydrolysis to TPA.

Manufacturing Processes

1 BROMINE-PROMOTED AIR OXIDATION OF *p*-XYLENE

Material Balance

Chemical	Coefficient T/T Product
Acetic acid	−0.06
Terephthalic acid (fiber grade)	1.00
p-Xylene	−0.67

Primary energy requirements for utilities   0.34 FOET/T
Unit investment for a 150 kT plant (1977 $)   0.91 $1000/T

Comments   Air oxidation of p-xylene occurs in liquid phase at 400°F and 200 psia in the presence of a cobalt-manganese-bromine catalyst.

2 PURIFICATION OF CRUDE TEREPHTHALIC ACID

Material Balance

Chemical	Coefficient T/T Product
Terephthalic acid (crude)	−1.02
Terephthalic acid (fiber grade)	1.00

Primary energy requirements for utilities   0.59 FOET/T
Unit investment for a 181 kT plant (1977 $)   0.15 $1000/T

3 OXIDATION OF p-XYLENE

Material Balance

Chemical	Coefficient T/T Product
Acetic acid	0.26
Methyl ethyl ketone	−0.24
Oxygen	−0.88
Terephthalic acid (crude)	1.00
p-Xylene	−0.68

Primary energy requirements for utilities   0.37 FOET/T
Unit investment for a 181 kT plant (1977 $)   0.24 $1000/T

4 REACTION OF p-XYLENE AND ACETALDEHYDE

Material Balance

Chemical	Coefficient T/T Product
Acetaldehyde	−0.63
Acetic acid	0.55
Terephthalic acid (crude)	1.00
p-Xylene	−0.72

Primary energy requirements for utilities   0.70 FOET/T
Unit investment for a 181 kT plant (1977 $)   0.21 $1000/T

TETRAHYDROFURAN

Tetrahydrofuran (THF) is consumed as a solvent for polyvinyl chloride, vinylidene chloride copolymers, and polyurethanes. It is used as a raw material for polytetramethylene glycol which is consumed in the production of spandex fibers and polyurethane elastomers. Small quantities of THF are also used as a solvent for cleaning kettles used in the production of PVC resins. A rapidly growing, though small, use for THF is in vinyl cements for joining PVC pipe parts. Most THF is produced by dehydrating 1,4-butanediol.

TOLUENE

The traditional markets for toluene, blending in gasoline and production of explosives (TNT), are now secondary to benzene production, which in 1970 accounted for over half the demand for toluene in the United States. Of the various other chemicals derived from toluene, benzoic acid and toluene diisocyanate are the most important. Toluene also has some use as a solvent.

Refinery reforming units are the major suppliers of toluene. Coal carbonization processes, once the primary source of toluene, still account for some of the toluene supply. Toluene also occurs as a by-product in ethylene manufacture, particularly when naphtha or gas oil is the feedstock.

TOLUENE DIAMINE

The only major market for toluene diamine is the production of toluene diisocyanate for making polyurethane plastics. The usual commercial product, an 80:20 mixture of the 2,4 and 2,6 isomers, is prepared by the catalytic hydrogenation of the corresponding mixture of dinitrotoluene isomers.

Manufacturing Processes

1 HYDROGENATION OF DINITROTOLUENE

Material Balance

Chemical	Coefficient T/T Product
Dinitrotoluene	−1.66
Hydrogen	−0.11
Toluene Diamine	1.00

Primary energy requirements for utilities   0.49 FOET/T
Unit investment for a 36 kT plant (1977 $)   0.42 $1000/T

Comments Reaction occurs at 500 psig and 165°F over a palladium-on-carbon catalyst.

TOLUENE DIISOCYANATE

Toluene diisocyanate is used primarily to produce polyurethanes, polymers which were first commercialized in 1955. The usual commercial product, an 80:20 mixture of 2,4- and 2,6-toluene diisocyanate, is manufactured by the phosgenation of the corresponding mixture of toluene diamine isomers.

Manufacturing Processes

1 PHOSGENATION OF TOLUENE DIAMINE

Material Balance

Chemical	Coefficient T/T Product
Chlorobenzene	−0.01
Phosgene	−1.26
Toluene diamine	−0.76
Toluene diisocyanate	1.00

Primary energy requirements for utilities   0.43 FOET/T
Unit investment for a 45 kT plant (1977 $)   0.43 $1000/T

Comments Reaction occurs at about 250°F using chlorobenzene as the solvent.

TRICHLOROETHYLENE

Trichloroethylene is used largely in the vapor degreasing of fabricated metal parts. It has various other uses as a solvent.

Today most trichloroethylene is derived from ethylene dichloride by chlorination or oxychlorination. In practice the chlorination and oxychlorination reactions are often made to occur simultaneously. In the original route to trichloroethylene, acetylene was chlorinated to 1,1,2,2-tetrachloroethane, which was dehydrochlorinated with lime. Later many manufacturers switched to a catalytic dehyrochlorination process, thereby avoiding the loss of chlorine as calcium chloride. In 1970 the acetylene-based processes were used to produce about half the trichloroethylene. Today, however, these processes are not commercially important in the United States.

TRIETHYLENE GLYCOL

Natural gas dehydration is the major market for triethylene glycol. It is also important as a humectant for tobacco, as a solvent for printing inks, and as a solvent for extracting aromatics from petroleum refinery streams. Polyester and polyurethane resins and vinyl plasticizers are important derivatives.

Triethylene glycol is produced as a by-product when ethylene oxide is hydrated to ethylene glycol. The quantities so produced have not been sufficient to satisfy the demand, so additional triethylene glycol is produced from the reaction of ethylene oxide and diethylene glycol. It is this reaction that is responsible for the formation of by-product triethylene glycol; the extent to which it occurs can be increased by decreasing the water-to-ethylene oxide ratio in the feed to the hydration process. However, this also increases the amount of by-product diethylene glycol, for which there may be no market. Thus the demand for additional triethylene glycol is not met by increasing by-product formation, but by reacting diethylene glycol and ethylene oxide in a separate process.

UREA

Most urea is consumed as a fertilizer or animal feed. Important nonagricultural markets include the production of melamine and the production of plastics such as urea-formaldehyde. Since the 1930s all urea manufactured in the United States has been derived from ammonia and carbon dioxide.

Manufacturing Processes

1 REACTION OF AMMONIA AND CARBON DIOXIDE

Material Balance

Chemical	Coefficient T/T Product
Ammonia	−0.57
Carbon dioxide	−0.74
Urea	1.00

Primary energy requirements for utilities	0.07 FOET/T
Unit investment for a 340 kT plant (1977 $)	0.16 $1000/T

2 TOTAL RECYCLE PROCESS

Material Balance

Chemical	Coefficient T/T Product
Ammonia	−0.57
Carbon dioxide	−0.74
Urea	1.00

Primary energy requirements for utilities	0.11 FOET/T
Unit investment for a 340 kT plant (1977 $)	0.19 $1000/T

VINYL ACETATE

Virtually all vinyl acetate is consumed in polymerized form as polyvinyl acetate, polyvinyl alcohol, polyvinyl butyral, and other polymers and copolymers. Polyvinyl acetate is used primarily in water-based paints and adhesives; polyvinyl alcohol is used in textile and paper processing; and polyvinyl butyral is used as an interlayer in safety glass.

For many years the usual route to vinyl acetate was the reaction of acetylene and acetic acid. Within the last decade, however, this process has been largely replaced by a process that combines ethylene and acetic acid. The ethylene route was first commercialized in 1966 and by 1970 accounted for more than 40% of the vinyl acetate capacity in the United States. Originally a liquid-phase process was used, but severe corrosion problems were encountered, forcing shutdown in favor of a vapor-phase process. Another route to vinyl acetate involves the reaction of acetaldehyde and acetic anhydride. This process was operated in the United States from 1953 to 1969.

Manufacturing Processes

1 REACTION OF ETHYLENE AND ACETIC ACID

Material Balance

Chemical	Coefficient T/T Product
Acetic acid	−0.70
Ethylene	−0.39
Oxygen	−0.33
Vinyl acetate	1.00

Primary energy requirements for utilities   0.39 FOET/T

Unit investment for a 136 kT plant (1977 $)   0.38 $1000/T

Comments   Reaction occurs in vapor phase at 360°F and 120 psia using a Pd catalyst.

2 REACTION OF ACETYLENE AND ACETIC ACID

Material Balance

Chemical	Coefficient T/T Product
Acetaldehyde	0.01
Acetic acid	−0.72
Acetylene	−0.32
Vinyl acetate	1.00

Primary energy requirements for utilities   0.15 FOET/T

Unit investment for a 136 kT plant (1977 $)   0.27 $1000/T

Comments   Reaction occurs at about 20 psia and 350–400°F using a zinc acetate-on-activated carbon catalyst.

3 REACTION OF ETHANE AND ACETIC ACID

Material Balance

Chemical	Coefficient T/T Product
Acetic acid	−0.76
Ethane	−0.44
Vinyl acetate	1.00
Miscellaneous chemicals	−0.02

Primary energy requirements for utilities   0.95 FOET/T

Unit investment for a 136 kT plant (1977 $)   0.35 $1000/T

VINYL CHLORIDE

Nearly all vinyl chloride is polymerized; polyvinyl chloride is the usual product, but copolymers such as vinyl chloride-vinyl acetate or vinyl chloride-vinylidene chloride (Saran) are also important.

Vinyl chloride is manufactured commercially from acetylene or from ethylene via ethylene dichloride. The acetylene route was once dominant but has gradually been replaced by the ethylene dichloride route.

Manufacturing Processes

1 CHLORINATION AND OXYCHLORINATION OF ETHYLENE

Material Balance

Chemical	Coefficient T/T Product
Chlorine	−0.61
Ethylene	−0.48
Sodium hydroxide	−0.01
Vinyl chloride	1.00

Primary energy requirements for utilities	0.44 FOET/T
Unit investment for a 272 kT plant (1977 $)	0.29 $1000/T

2 DEHYDROCHLORINATION OF ETHYLENE DICHLORIDE

Material Balance

Chemical	Coefficient T/T Product
Ethylene dichloride	−1.66
Hydrogen chloride	0.61
Vinyl chloride	1.00

Primary energy requirements for utilities	0.17 FOET/T
Unit investment for a 181 kT plant (1977 $)	0.08 $1000/T

3 HYDROCHLORINATION OF ACETYLENE

Material Balance

Chemical	Coefficient T/T Product
Acetylene	−0.43
Hydrogen chloride	−0.60
Sodium hydroxide	−0.01
Vinyl chloride	1.00

Primary energy requirements for utilities	0.05 FOET/T
Unit investment for a 272 kT plant (1977 $)	0.13 $1000/T

VINYLIDENE CHLORIDE

Vinylidene chloride is copolymerized with vinyl chloride, and acrylic and methacrylic acid esters to form products which are used in film production (Saran) with very large quantities being used for filament manufacture, latex production, and plastic-lined pipe.

Vinylidene chloride is produced by chlorination of ethylene dichloride to form 1,1,2-trichloroethane which in turn is dehydrochlorinated to vinylidene chloride. Ethane or vinyl chloride is also used instead of ethylene dichloride, but chlorination results in 1,1,1-trichloroethane (methyl chloroform).

Manufacturing Processes

1 DEHYDROCHLORINATION OF 1,1,2-TRICHLOROETHANE

Material Balance

Chemical	Coefficient T/T Product
Chlorine	−1.03
Ethylene	−0.09
Ethylene dichloride	−0.83
Hydrogen chloride	0.43
Sodium hydroxide	−0.46
Vinylidene chloride	1.00
By-product credits	0.13

Primary energy requirements for utilities	0.21 FOET/T
Unit investment for a 23 kT plant (1977 $)	0.63 $1000/T

2 FROM VINYL CHLORIDE BY DEHYDROCHLORINATION OF METHYL CHLOROFORM

Material Balance

Chemical	Coefficient T/T Product
Chlorine	−0.92
Hydrogen chloride	0.47
Trichloroethylene	0.13
Vinyl chloride	−0.72
Vinylidene chloride	1.00
By-product credits	0.04
Miscellaneous chemicals & catalyst	−0.01

Primary energy requirements for utilities   0.35 FOET/T
Unit investment for a 23 kT plant (1977 $)   0.91 $1000/T

Comments   Vinyl chloride is hydrochlorinated to 1,1-dichloroethane in a ferric chloride catalyzed liquid phase reaction. The 1,1-dichloroethane is thermally chlorinated at 896°F and 85 psia to methyl chloroform, vinylidene chloride, and vinyl chloride. The methyl chloroform is thermally cracked at 950°F and 85 psia to vinylidene chloride.

3   ETHANE CHLORINATION AND THERMAL DEHYDROCHLORINATION OF METHYL CHLOROFORM

Material Balance

Chemical	Coefficient T/T Product
Chlorine	−3.01
Ethane	−0.56
Ethyl chloride	0.08
Hydrogen chloride	2.11
Vinylidene chloride	1.00
By-product credits	0.21

Primary energy requirements for utilities   0.62 FOET/T
Unit investment for a 23 kT plant (1977 $)   1.10 $1000/T

m-XYLENE

The only large-scale use for pure *m*-xylene is the production of isophthalic acid. Mixed xylenes have various solvent uses. Refinery reforming operations supply most of the xylene consumed by the petrochemical industry. *m*-Xylene is the most abundant isomer in reformate but is also the least demanded. Thus, it is usually isomerized to a mixture of the other isomers; this is usually an integral part of isomer separation plants.

o-XYLENE

The only large scale use of pure *o*-xylene is the production of phthalic anhydride. Mixed xylenes have various solvent uses. Refinery reforming operations supply most of the xylene consumed by the petrochemical industry. *m*-Xylene is the most abundant isomer in reformate but is also the least demanded. Thus, additional *o*-xylene is usually produced from *m*-xylene by isomerization; this operation is usually an integral part of isomer separation plants.

p-XYLENE

The only large-scale use for p-xylene is the production of terephthalic acid or its dimethyl ester. Refinery reforming operations supply most of the xylene consumed by the petrochemical industry. m-Xylene is the most abundant isomer in reformate but is also the least demanded. Thus, additional p-xylene is usually produced from m-xylene by isomerization; this operation is usually an integral part of isomer separation plants.

Manufacturing Processes

1 ISOMERIZATION OF m-XYLENE (AROMAX-ISOLENE)

Material Balance

Chemical	Coefficient T/T Product
p-Xylene	1.00
Xylenes	−1.15

Primary energy requirements for utilities   0.65 FOET/T
Unit investment for a 91 kT plant (1977 $)   0.28 $1000/T

2 ISOMERIZATION OF m-XYLENE (PAREX-ISOMAR)

Material Balance

Chemical	Coefficient T/T Product
Hydrogen	−0.01
p-Xylene	1.00
Xylenes	−1.25

Primary energy requirements for utilities   0.56 FOET/T
Unit investment for a 91 kT plant (1977 $)   0.21 $1000/T

XYLENES

The largest outlet for xylenes presently is gasoline octane improvement. Motor gasoline will probably continue to be the largest use, because the xylenes-rich fraction left behind after benzene and toluene are removed from reformates is returned to the gasoline blending pool. Considerable quantities of mixed xylenes have also found use as solvents.

Methods of producing xylenes follow closely those used for toluene. Xylenes are obtained from certain petroleum fractions by catalytic reforming or hydroforming. The catalytic reforming of a selected naphtha cut yields a reformate rich in benzene, toluene, and xylenes. By a combination of extraction and distillation processes, a mixed xylenes fraction is produced. Xylenes are also derived from processes involving the disproportionation of toluene or the trans-alkylation of toluene with trimethylbenzenes. The products are principally benzene and xylenes.

CHAPTER SEVEN

Plastics and Resins

ACETAL RESINS

Acetal resins are part of the family of thermoplastic engineering resins which are finding increasing use as replacements for fabricated metal parts in a variety of applications. Acetal resins are fabricated mostly by injection molding and the resulting products have found their way in areas such as transportation, plumbing, appliances, machinery/industrial, electrical/electronics, and others.

These resins are homo- and/or copolymers of formaldehyde. They are characterized by their good electrical and mechanical properties, and strong resistance to high temperature, moisture, and organic solvents.

Manufacturing Processes

1 ACETAL RESIN FROM TRIOXANE AND ETHYLENE OXIDE

Material Balance

Chemical	Coefficient T/T Product
Acetal resin	1.00
Ammonia	−0.10
Cyclohexane	−0.01
Ethylene oxide	−0.08
Methanol	−1.48
Sulfuric acid	−0.04
Miscellaneous chemicals and catalyst	−0.02

Primary energy requirements for utilities   1.32 FOET/T
Unit investment for an 18 kT plant (1977 $)   2.20 $1000/T

Comments   Formaldehyde is trimerized in sulfuric acid at 102°C to form trioxane. Trioxane and ethylene oxide are copolymerized in cyclohexane, using boron-trifluoride dietherate catalyst. The solid polymer is reslurried in ammonia and heated to 300–320°F.

2   ACETAL RESIN FROM FORMALDEHYDE

Material Balance

Chemical	Coefficient T/T Product
Acetal resin	1.00
Acetic anhydride	−0.10
Acetone	−0.03
Cyclohexane	−0.01
Cyclohexanol	−0.01
Methanol	−1.32
Miscellaneous chemicals	−0.01

Primary energy requirements for utilities   4.08 FOET/T
Unit investment for an 18 kT plant (1977 $)   2.97 $1000/T

ACRYLONITRILE-BUTADIENE-STYRENE (ABS)

Acrylonitrile-butadiene-styrene (ABS) is a class of polymers with over 50% styrene and varying amounts of the other two compounds. ABS resins are two-phase systems. Polybutadiene is dispersed in the rigid styrene-acrylonitrile copolymer, resulting in thermoplastic materials of varying degree of hardness and toughness.

Large-volume applications of ABS resins include pipe and pipe fittings, accounting for about 28% of these resins' total consumption in 1977. The second largest market for this material is in automotive components, which take about 20% of its total outlet. They are also used in appliance components—14%, components of business machines, telephones, and electrical and electronic equipment—10%, recreational vehicle components—7%. Other uses such as packaging, luggage and cases, toys, and sporting goods, consume the rest of ABS resins' outlet.

Manufacturing Processes

1 ABS GRAFT RESIN BY EMULSION/EMULSION POLYMERIZATION

Material Balance

Chemical	Coefficient T/T Product
ABS resin	1.00
Acrylonitrile	−0.19
Butadiene	−0.25
Styrene	−0.54
Miscellaneous chemicals	−0.05

Primary energy requirements for utilities	0.21 FOET/T
Unit investment for a 55 kT plant (1977 $)	0.95 $1000/T

2 ABS GRAFT RESIN BY SUSPENSION/EMULSION POLYMERIZATION

Material Balance

Chemical	Coefficient T/T Product
ABS resin	1.00
Acrylonitrile	−0.19
Butadiene	−0.25
Styrene	−0.54
Miscellaneous chemicals	−0.05

Primary energy requirements for utilities	0.20 FOET/T
Unit investment for a 55 kT plant (1977 $)	0.89 $1000/T

3 ABS GRAFT RESIN BY BULK/SUSPENSION POLYMERIZATION

Material Balance

Chemical	Coefficient T/T Product
ABS resin	1.00
Acrylonitrile	−0.22
Polybutadiene	−0.07
Styrene	−0.67
Miscellaneous chemicals	−0.04

Primary energy requirements for utilities	0.19 FOET/T
Unit investment for a 55 kT plant (1977 $)	0.71 $1000/T

BARRIER RESINS

Commercially, there are two different types of barrier resins: nitrile barrier resins, which are products of polymerization of acrylonitrile with other monomers such as acrylates, butadiene, and styrene; and polyethylene terephthalate (PET) barrier resin, which is produced by polycondensation of ethylene glycol with either dimethyl terephthalate or terephthalic acid. Nitrile resins generally offer carbon dioxide barrier properties better than polyvinyl chloride and polypropylene. However, nonpolar resins such as polyolefins have better water vapor barrier properties than nitrile resins. The gas and water vapor barrier performance of PET barrier resin is intermediate between that of the nitriles and polyolefins. The food packaging industry, therefore, is a potential user of these resins. Nitrile and PET barrier resins are in commercial use in the form of blow molded bottles and jars, ranging in size from about one half ounce to 64 ounces. They have also been used in the form of film and sheet in wrapping cheese slices, cured meats, and other foodstuffs.

Manufacturing Processes

1 NITRILE BARRIER RESIN (BAREX) BY EMULSION POLYMERIZATION

Material Balance

Chemical	Coefficient T/T Product
Acrylonitrile	−0.73
Barrier resin	1.00
Butadiene	−0.10
Methyl acrylate	−0.23
Miscellaneous chemicals and catalyst	−0.05

Primary energy requirements for utilities	0.19 FOET/T
Unit investment for a 27 kT plant (1977 $)	0.81 $1000/T

2 NITRILE BARRIER RESIN (CYCOPAC) BY EMULSION POLYMERIZATION

Material Balance

Chemical	Coefficient T/T Product
Acrylonitrile	−0.57
Barrier resin	1.00
Methyl methacrylate	−0.04
Styrene	−0.23
Styrene-butadiene rubber (latex)	−0.21

Primary energy requirements for utilities   0.23 FOET/T
Unit investment for a 27 kT plant (1977 $)   0.79 $1000/T

3 NITRILE BARRIER RESIN (LOPAC) BY SUSPENSION POLYMERIZATION

Material Balance

Chemical	Coefficient T/T Product
Acrylonitrile	−0.79
Barrier resin	1.00
Styrene	−0.26

Primary energy requirements for utilities   0.14 FOET/T
Unit investment for a 27 kT plant (1977 $)   0.72 $1000/T

EPOXY RESINS

The largest market for epoxy resins is in protective coatings such as auto primers, can and drum coatings, and pipe coatings. This use area accounted for about 50% of the total epoxy resins' outlet in 1978. Reinforced uses such as electrical laminates and filament winding were the second largest outlet for these materials, sharing about 18% of the resins' total domestic consumption. The other use areas were mainly tooling, casting, and molding—9%, bonding and adhesives—8%, flooring and paving—6%.

The first epoxy resins developed on a commercial scale are those made from epichlorohydrin and bisphenol-A. By varying the ratio of epichlorohydrin and bisphenol-A, as well as the operating conditions, resins of low, medium, and high molecular weight can be produced.

Manufacturing Processes

1 LOW-MOLECULAR-WEIGHT LIQUID DGEBA* RESIN BY BATCH PROCESS

Material Balance

Chemical	Coefficient T/T Product
Bisphenol-A	−0.67
Epichlorohydrin	−0.56
Epoxy resin	1.00
Methyl isobutyl ketone	−0.05
Sodium hydroxide	−0.24
Miscellaneous chemicals	−0.02

Primary energy requirements for utilities   0.57 FOET/T
Unit investment for a 9 kT plant (1977 $)   0.74 $1000/T

2 LOW-MOLECULAR-WEIGHT LIQUID DGEBA* RESIN BY CONTINUOUS PROCESS

Material Balance

Chemical	Coefficient T/T Product
Bisphenol-A	−0.74
Epichlorohydrin	−0.56
Epoxy resin	1.00
Ethanol	−0.05
Sodium hydroxide	−0.25

Primary energy requirements for utilities   0.26 FOET/T
Unit investment for a 9 kT plant (1977 $)   0.54 $1000/T

3 SOLID DGEBA* RESIN BY BATCH PROCESS

Material Balance

Chemical	Coefficient T/T Product
Bisphenol-A	−0.76
Epichlorohydrin	−0.39
Epoxy resin	1.00
Methyl isobutyl ketone	−0.04
Sodium hydroxide	−0.20
Miscellaneous chemicals	−0.03

Primary energy requirements for utilities   0.26 FOET/T
Unit investment for a 9 kT plant (1977 $)   0.76 $1000/T

MELAMINE-FORMALDEHYDE RESINS

Melamine-formaldehyde accounts for about 20% of all the amino resins produced in the United States. This material is used in areas such as laminating, paper treating and coating, protective coating, and textile treating and coating.

*Diglycidyl ethers of bisphenol-A.

Laminating uses for melamine resins are in the manufacture of tabletops, countertops, and wall paneling; they have also gained importance in fire-resistant industrial and electrical applications. Melamine-formaldehyde resins are also used as molding powder in the manufacture of dinnerware, buttons, sanitaryware, and so forth.

Manufacturing Processes

1 MELAMINE-FORMALDEHYDE MOLDING COMPOUND

Material Balance

Chemical	Coefficient T/T Product
Formaldehyde	−0.23
Melamine	−0.48
Melamine-formaldehyde	1.00
Miscellaneous chemicals	−0.28

Primary energy requirements for utilities	0.24 FOET/T
Unit investment for a 2 kT plant (1977 $)	1.13 $1000/T

2 MELAMINE-FORMALDEHYDE RESIN SYRUP

Material Balance

Chemical	Coefficient T/T Product
Formaldehyde	−0.33
Melamine	−0.70
Melamine-formaldehyde	1.00
Miscellaneous chemicals	−0.01

Primary energy requirements for utilities	0.06 FOET/T
Unit investment for a 2 kT plant (1977 $)	0.19 $1000/T

NYLON RESINS

Nylon resins belong to a group of high-performance plastics known as engineering thermoplastics. They are converted to end products mostly by injection molding, and to a smaller extent, by extrusion. In 1977, injection molding resins were used in areas such as appliances—6%, consumer products—14%, electrical/electronics—14%, industrial—11%, and transportation—26%. Extrusion resins were mainly consumed in nontextile filaments—9%, film—8%, sheet, rod, and tube—5%, and wire and cable—3%.

Nylon 6 and 66 are the most important types of nylon resins produced in the United States. Nylon 66 is the polymer of hexamethylenediamine and adipic acid and generally offers somewhat higher strength values, greater hardness and stiffness, and lower elongation than nylon 6 which is made from caprolactam.

Manufacturing Processes

1 NYLON 6 CHIPS FROM CAPROLACTAM

Material Balance

Chemical	Coefficient T/T Product
Caprolactam	−0.95
Nylon 6	1.00
Miscellaneous chemicals	−0.09

Primary energy requirements for utilities	0.33 FOET/T
Unit investment for a 15 kT plant (1977 $)	0.51 $1000/T

Comments Caprolactam is polymerized continuously for 18 hours at 500°F. A small amount of water is used as a catalyst.

2 NYLON 6 MELT FROM CAPROLACTAM

Material Balance

Chemical	Coefficient T/T Product
Caprolactam	−0.95
Nylon 6	1.00
Miscellaneous chemicals	−0.09

Primary energy requirements for utilities	0.14 FOET/T
Unit investment for a 15 kT plant (1977 $)	0.39 $1000/T

3 NYLON 66 CHIPS FROM ADIPIC ACID AND HEXAMETHYLENEDIAMINE

Material Balance

Chemical	Coefficient T/T Product
Active carbon	−0.02
Adipic acid	−0.65
Hexamethylenediamine	−0.52
Methanol	−0.01
Nylon 66	1.00

Primary energy requirements for utilities	0.24 FOET/T
Unit investment for a 14 kT plant (1977 $)	0.67 $1000/T

Comments Nylon-salt is made first by reacting hexamethylenediamine and adipic acid in methanol. Nylon 66 is formed by polycondensation of nylon salt in a batch autoclave on a 5-hour cycle operation.

PHENOL-FORMALDEHYDE RESINS

Phenolic resins are the oldest synthetic plastics materials. These materials were developed in 1909 and offer outstanding strength and hardness, electrical properties, and temperature resistance. Approximately two thirds of domestic consumption of phenolic resins are used in bonding and adhesives, laminating, plywood, and protective coatings. Molding compounds account for the remainder of the resins which are used in areas such as appliances, automotive, closures, housewares, and electrical/electronics.

More than 95% of phenolic resins are based on phenol and formaldehyde monomers. However, small amounts of phenolic resins involve other coreactants such as o-cresol and furfural as replacements for phenol and formaldehyde, respectively.

Manufacturing Processes

1 PHENOL-FORMALDEHYDE NOVOLAC MOLDING POWDER

Material Balance

Chemical	Coefficient T/T Product
Formaldehyde	−0.25
Hexamethylenetetramine	−0.10
Phenol	−0.90
Phenol-formaldehyde	1.00
Miscellaneous and filler	−1.02

Primary energy requirements for utilities   0.11 FOET/T
Unit investment for a 7 kT plant (1977 $)   0.68 $1000/T

Comments   Reaction of phenol and formaldehyde is carried out at 212°F for 2 hours.

2 PHENOL-FORMALDEHYDE RESOL SYRUP BY BATCH PROCESS

Material Balance

Chemical	Coefficient T/T Product
Formaldehyde	−0.34
Phenol	−0.83
Phenol-formaldehyde	1.00
Sodium hydroxide	−0.03
Sulfuric acid	−0.04
Miscellaneous chemicals	−0.01

Primary energy requirements for utilities   0.01 FOET/T
Unit investment for a 25 kT plant (1977 $)   0.24 $1000/T

Comments   Reaction of phenol and formaldehyde is catalyzed by sodium hydroxide at 140°F.

POLYACRYLAMIDE

The largest market for polyacrylamide in the United States is in water and effluent treatment which consumes over 50% of polyacrylamide's total outlet. It is also used in other areas such as mining, oil production, coatings and adhesives, and textiles.

Manufacturing Processes

1 POLYACRYLAMIDE (ANIONIC) BY POLYMERIZATION AND HYDROLYSIS

Material Balance

Chemical	Coefficient T/T Product
Acrylamide	−0.96
Polyacrylamide	1.00
Sodium carbonate	−0.33

Primary energy requirements for utilities   0.81 FOET/T
Unit investment for a 5 kT plant (1977 $)   1.93 $1000/T

Comments Acrylamide is prepolymerized with potassium persulfate and sodium bisulfite as initiator at 112°F. Polymerization then continues in batch reactors at 195°F. Total polymerization time is 2 hours.

POLYACRYLATES

One of the largest users of polyacrylates is the paint industry. Polyacrylates are widely used as dispersions in water-based paints, particularly in exterior coatings. They are also used in products such as automotive topcoats and primers, coil coatings, and appliances.

Polyacrylates are polymers of esters of acrylic acid and methacrylic acid. Since a number of such esters can be copolymerized with each other and also modified with a variety of nonacrylic monomers, there are hundreds of polyacrylates available. However, polymethyl methacrylate is one of the most important ones which is widely used in glazing, lighting fixtures, and panels and sidings.

Manufacturing Processes

1 POLYACRYLATE LATEX (47‰ POLYMER) BY BATCH EMULSION POLYMERIZATION

Material Balance

Chemical	Coefficient T/T Product
Ethyl acrylate	−0.64
Methacrylic acid	−0.01
Methyl methacrylate	−0.31
Polyacrylate	1.00
Miscellaneous chemicals	−0.08

Primary energy requirements for utilities   0.19 FOET/T
Unit investment for a 7 kT plant (1977 $)   0.70 $1000/T

2 POLYACRYLATE (PELLETS) BY BATCH SUSPENSION POLYMERIZATION

Material Balance

Chemical	Coefficient T/T Product
Ethyl acrylate	−0.16
Hydrogen chloride	−0.01
Methyl methacrylate	−0.90
Polyacrylate	1.00
Miscellaneous chemicals	−0.06

Primary energy requirements for utilities	0.25 FOET/T
Unit investment for a 7 kT plant (1977 $)	0.84 $1000/T

Comments Polymerization takes place at 230°F and 3 atm. Polymerization time is 1¹/₂ hour, and cooling time, 1 hour.

3 POLYMETHYL METHACRYLATE SHEET BY CONTINUOUS CASTING

Material Balance

Chemical	Coefficient T/T Product
Methyl methacrylate	−1.00
Polymethyl methacrylate	1.00

Primary energy requirements for utilities	0.19 FOET/T
Unit investment for an 18 kT plant (1977 $)	1.63 $1000/T

4 POLYMETHYL METHACRYLATE SHEET BY BATCH CASTING

Material Balance

Chemical	Coefficient T/T Product
Methyl methacrylate	−1.00
Polymethyl methacrylate	1.00

Primary energy requirements for utilities	0.07 FOET/T
Unit investment for an 18 kT plant (1977 $)	0.52 $1000/T

POLYBUTENES AND POLYISOBUTYLENES

Commercialization of butylene polymers started in 1933 with polymerizing highly purified isobutylene followed by polybutenes' production just before World War II. Polybutenes are relatively low-molecular-weight polymers with properties very similar to low-molecular-weight polyisobutylenes. These materials are typically used in adhesives, caulking compounds, sealants, and coatings, or as additives for internal combustion engine oil, and plasticizers.

Manufacturing Processes

1 POLYBUTENES FROM BUTENES

Material Balance

Chemical	Coefficient T/T Product
Aluminum chloride	−0.02
Butane	−0.50
Butenes	−0.57
Polybutenes	1.00
Sodium hydroxide	−0.09

Primary energy requirements for utilities	0.17 FOET/T
Unit investment for a 27 kT plant (1977 $)	0.38 $1000/T

2 POLYISOBUTYLENES FROM ISOBUTYLENE

Material Balance

Chemical	Coefficient T/T Product
Aluminum chloride	−0.01
n-Hexane	−0.10
Isobutylene	−1.01
Polyisobutylenes	1.00
Sodium hydroxide	−0.01

Primary energy requirements for utilities	0.66 FOET/T
Unit investment for an 11 kT plant (1977 $)	0.71 $1000/T

POLYBUTYLENE TEREPHTHALATE (PBT) RESINS

The largest markets for PBT resins are in automotive and truck components and parts, which take about 42% of PBT's total outlet; electrical and electronic components consume 32%, and the remainder is mainly used in industrial and appliance components and parts.

PBT resins are highly crystalline engineering thermoplastic resins made from approximately equal molar proportions of 1,4-butanediol and dimethyl terephthalate. About 80% of all PBT resins are filled, and more than 95% of this amount are glass-filled. Outstanding mechanical properties of these compounds make them competitive with the other synthetic resins, primarily engineering thermoplastics such as nylons, polyacetal, and polycarbonates.

Manufacturing Processes

1 FLAME RETARDANT PBT RESIN (GLASS-FILLED)

Material Balance

Chemical	Coefficient T/T Product
1,4-Butanediol	−0.28
Dimethyl terephthalate	−0.56
Methanol	0.18
PBT resin	1.00
Miscellaneous chemicals (including glass fiber)	−0.37

Primary energy requirements for utilities	0.28 FOET/T
Unit investment for a 15 kT plant (1977 $)	1.37 $1000/T

2 PLAIN PBT RESIN

Material Balance

Chemical	Coefficient T/T Product
1,4-Butanediol	−0.44
Dimethyl terephthalate	−0.89
Methanol	0.29
PBT resin	1.00

Primary energy requirements for utilities	0.24 FOET/T
Unit investment for a 9 kT plant (1977 $)	1.90 $1000/T

POLYCARBONATE RESINS

Polycarbonate resins are among engineering thermoplastics which are used in several specialty applications. Major markets for these resins include safety and antivandalism glazing, appliances, power tools, communications equipment, business machines, lighting, signs, and automotive taillights. Polycarbonate resins are characterized by their high impact strength, excellent electrical properties and dimensional stability, high heat distortion temperature, and good optical clarity. Production of these materials, however, is limited because of their relatively poor abrasion and solvent resistance, high cost, and also because of a variety of other engineering thermoplastics that can complete favorably in many applications.

Commercially, polycarbonate resins are derived from the reaction of phosgene and bisphenol-A by interfacial or solution phosgenation.

Manufacturing Processes

1 POLYCARBONATE RESIN BY CONTINUOUS SOLUTION PHOSGENATION

Material Balance

Chemical	Coefficient T/T Product
Bisphenol-A	−0.91
Heptane	−0.02
Hydrogen chloride	−0.18
Methylene chloride	−0.01
Phosgene	−0.42
Polycarbonate	1.00
Sodium hydroxide	−0.54
Miscellaneous chemicals	−0.02

Primary energy requirements for utilities	1.11 FOET/T
Unit investment for an 18 kT plant (1977 $)	2.05 $1000/T

2 POLYCARBONATE RESIN BY INTERFACIAL PHOSGENATION

Material Balance

Chemical	Coefficient T/T Product
Bisphenol-A	−0.91
Hydrogen chloride	−0.03
Methylene chloride	−0.01
Phosgene	−0.44
Polycarbonate	1.00
Sodium hydroxide	−0.39
Miscellaneous chemicals	−0.02

Primary energy requirements for utilities   0.68 FOET/T
Unit investment for a 9 kT plant (1977 $)   2.11 $1000/T

3 POLYCARBONATE RESIN BY BATCH SOLUTION PHOSGENATION

Material Balance

Chemical	Coefficient T/T Product
Bisphenol-A	−0.91
Heptane	−0.02
Hydrogen chloride	−0.18
Methylene chloride	−0.01
Phosgene	−0.42
Polycarbonate	1.00
Sodium hydroxide	−0.54
Miscellaneous chemicals	−0.02

Primary energy requirements for utilities   1.11 FOET/T
Unit investment for an 18 kT plant (1977 $)   2.09 $1000/T

4 POLYCARBONATE RESIN (FLAME RESISTANT GRADE)

Material Balance

Chemical	Coefficient T/T Product
Bisphenol-A	−0.82
Heptane	−0.02
Hydrogen chloride	−0.17
Methylene chloride	−0.01
Phosgene	−0.40
Polycarbonate	1.00
Sodium hydroxide	−0.51
Miscellaneous chemicals	−0.11

Primary energy requirements for utilities   1.11 FOET/T
Unit investment for an 18 kT plant (1977 $)   2.01 $1000/T

POLYETHER POLYOLS

Polyether polyols are used in two different applications: (1) urethane applications, (2) nonurethane applications. About 80% of the total domestic consumption of polyether polyols is in urethane applications to manufacture mainly flexible and rigid foams. Among the urethane polyols, glycerine adduct is the leading polyol followed by polypropylene glycol and other propylene oxide based adducts. Nonurethane polyols include polypropylene glycol, polyethylene glycol and other ethylene oxide-based adducts, and propylene oxide-ethylene oxide random and block copolymers. Nonurethane polyols are used mainly as surface active agents and as lubricants and functional fluids. As surface active agents, they are used as defoamers and antifoam agents, and as wetting agents in oil well acidizing with dilute hydrochloric acid. They are also used in low-sudsing synthetic detergents, fermentation processes, steam generating systems, cosmetics and pharmaceuticals, and as additives in automatic dishwasher rinses. In their second largest market, as lubricants and functional fluids, they are used in automotive brake fluids and hydraulic fluids, textile fiber lubricants, heat transfer fluid, metal working fluids, and compressor lubricants.

Manufacturing Processes

1 POLYETHER POLYOL (SORBITOL-BASED HEXOL)

Material Balance

Chemical	Coefficient T/T Product
Polyether polyol	1.00
Propylene oxide	−0.78
Sorbitol	−0.23

Primary energy requirements for utilities	0.12 FOET/T
Unit investment for a 45 kT plant (1977 $)	0.33 $1000/T

Comments Propylene oxide is added to the mixture of sorbitol and water at 230°F. After removing water, the reaction of sorbitol and propylene oxide is continued.

2 POLYETHER POLYOL (PHOSPHORUS-CONTAINING)

Material Balance

Chemical	Coefficient T/T Product
Glycerine	−0.21
Phosphoric acid	−0.23
Polyether polyol	1.00
Propylene oxide	−0.65

Primary energy requirements for utilities   0.05 FOET/T
Unit investment for a 5 kT plant (1977 $)   0.89 $1000/T

Comments Glycerine and phosphoric acid are reacted at 112°F for 2 hours. After removing water at 260°F, propylene oxide is added to the mixture at 176°F.

3 POLYETHER POLYOL (GLYCERINE-BASED TRIOL)

Material Balance

Chemical	Coefficient T/T Product
Glycerine	−0.03
Polyether polyol	1.00
Propylene oxide	−0.98

Primary energy requirements for utilities   0.18 FOET/T
Unit investment for a 45 kT plant (1977 $)   0.33 $1000/T

4 POLYETHYLENE GLYCOL FROM ETHYLENE OXIDE AND DIETHYLENE GLYCOL

Material Balance

Chemical	Coefficient T/T Product
Diethylene glycol	−0.27
Ethylene oxide	−0.75
Phosphoric acid	−0.01
Polyethylene glycol	1.00
Sodium hydroxide	−0.02

Primary energy requirements for utilities   0.01 FOET/T
Unit investment for a 9 kT plant (1977 $)   0.33 $1000/T

5 POLYPROPYLENE GLYCOL FROM PROPYLENE OXIDE

Material Balance

Chemical	Coefficient T/T Product
Active carbon	−0.02
Polypropylene glycol	1.00
Propylene glycol	−0.06
Propylene oxide	−0.98
Sulfuric acid	−0.01
Miscellaneous chemicals	−0.02

Primary energy requirements for utilities   0.09 FOET/T
Unit investment for a 9 kT plant (1977 $)   0.42 $1000/T

Comments   Propylene oxide is added to propylene glycol at 250°F and 2 atm using potassium hydroxide as catalyst.

POLYETHYLENE (HIGH DENSITY)

High-density polyethylene (HDPE) resins have become one of the major thermoplastics within a relatively short time. Blow-molded products form the largest market for HDPE resins primarily in bottles for household chemicals and dairy bottles. Extruded products are the second largest outlet for HDPE resins. They are mainly used in pipe and tubing, primarily in corrugated drainage tubing. They are also used in wire and cable, and as film in production of heavy-duty shipping bags. Injection molding is the last major category for the products of HDPE resins. In this area, they are used in manufacture of such products as food containers, shipping pails, housewares, toys, and closures and caps. Polypropylene has made inroads in the HDPE injection molding resin market because of their favorable economics. Housewares, toys, closures, and furniture parts are the areas where polypropylene has replaced HDPE to some extent.

Manufacturing Processes

1 HDPE BY PHILLIPS TECHNOLOGY

Material Balance

Chemical	Coefficient T/T Product
Butene-1	−0.01
Ethylene	−1.02
HDPE	1.00
Pentane	−0.05

Primary energy requirements for utilities   0.15 FOET/T
Unit investment for a 91 kT plant (1977 $)   0.44 $1000/T

Comments   The polymerization is carried out in *n*-pentane at 210°F and 35 atm using chromia-silica catalyst.

2 HDPE BY SOLVAY TECHNOLOGY (ZIEGLER CATALYST)

Material Balance

Chemical	Coefficient T/T Product
Butene-1	−0.02
Ethylene	−1.02
HDPE	1.00
Hexane	−0.03

Primary energy requirements for utilities   0.42 FOET/T
Unit investment for a 91 kT plant (1977 $)   0.53 $1000/T

3 HDPE BY UNION CARBIDE TECHNOLOGY (GAS PHASE PROCESS)

Material Balance

Chemical	Coefficient T/T Product
Ethylene	−1.02
HDPE	1.00

Primary energy requirements for utilities   0.19 FOET/T
Unit investment for a 91 kT plant (1977 $)   0.44 $1000/T

4 HDPE BY HOECHST TECHNOLOGY

Material Balance

Chemical	Coefficient T/T Product
Ethylene	−1.04
HDPE	1.00

Primary energy requirements for utilities   0.16 FOET/T
Unit investment for a 91 kT plant (1977 $)   0.44 $1000/T

Comments The polymerization takes place at 185°F and 7.8 atm using modified Ziegler catalysts (titanium-aluminum compounds) supported on oxygenated magnesium compounds.

5 HDPE BY MONTEDISON TECHNOLOGY

Material Balance

Chemical	Coefficient T/T Product
Ethylene	-1.02
HDPE	1.00
Hexane	-0.03
Propylene (polymer grade)	-0.02

Primary energy requirements for utilities	0.40 FOET/T
Unit investment for a 91 kT plant (1977 $)	0.53 $1000/T

Comments The reaction takes place at 185°F and 13 atm using a Ziegler catalyst (titanium-aluminum compound) on anhydrous magnesium chloride as the support.

6 HDPE BY STAMICARBON TECHNOLOGY

Material Balance

Chemical	Coefficient T/T Product
Ethylene	1.02
HDPE	1.00
Hexane	-0.04
Propylene (polymer grade)	-0.02

Primary energy requirements for utilities	0.25 FOET/T
Unit investment for a 91 kT plant (1977 $)	0.42 $1000/T

Comments The solution-form Ziegler catalyst components are fed to the reactor as separate solutions of titanium tetrachloride, dibutyl magnesium, and ethyl aluminum sesquichloride. Ethylene and propylene comonomers are absorbed in hexane, cooled to $-40°F$ to absorb the heat of polymerization and fed to the reactor which operates adiabatically at 265°F and 30 atm.

POLYETHYLENE (LOW DENSITY)

Low-density polyethylene (LDPE) resins are the volume leader of the thermoplastic resins. They are converted to end products primarily through extrusion processes and injection molding. LDPE injection molding resin market is mainly in housewares, lids, and toys and novelties. The largest use area for LDPE resins, however, is in the extruded film market which alone accounts for over 60% of the total resins' outlet. More than one-half of the LDPE film is used in food and nonfood packaging applications mainly in baked goods and industrial liners. The other food packaging applications of LDPE film include packaging candy, frozen food, and meat and poultry, and the other nonfood packaging users are garment bags, shipping sacks, heavy-duty bags, stretch pallet wraps, and shrinkable pallet wraps. The remainder of LDPE film is used in nonpackaging applications such as agriculture, construction, disposables, household, trash bags, and rubber industry.

Low-density polyethylene resins are generally produced by high-pressure processes. Two types of reactor are used, one is a continuous-flow, stirred autoclave, and the other is a tubular reactor. Generally, pure ethylene is charged to the reactor and no solvent is used.

Manufacturing Processes

1 LDPE BY COMPARTMENTED AUTOCLAVE REACTOR

Material Balance

Chemical	Coefficient T/T Product
Ethylene	−1.03
LDPE	1.00
Miscellaneous chemicals	−0.01

Primary energy requirements for utilities	0.34 FOET/T
Unit investment for a 100 kT plant (1977 $)	0.89 $1000/T

2 LDPE BY BACKMIXED AUTOCLAVE REACTOR

Material Balance

Chemical	Coefficient T/T Product
Ethylene	−1.03
LDPE	1.00
Miscellaneous chemicals	−0.01

Primary energy requirements for utilities	0.25 FOET/T
Unit investment for a 100 kT plant (1977 $)	0.77 $1000/T

3 LDPE BY TUBULAR REACTOR

Material Balance

Chemical	Coefficient T/T Product
Butane	−0.01
Ethylene	−1.03
LDPE	1.00

Primary energy requirements for utilities	0.34 FOET/T
Unit investment for a 100 kT plant (1977 $)	0.80 $1000/T

POLYETHYLENE TEREPHTHALATE RESINS

Polyethylene terephthalate (PET) resins are mainly used in the film market. Photographic film is the largest market for PET film which includes microfilm, amateur film, graphic arts film, x-ray film, industrial film, professional still film, and professional motion picture films. PET film is also used in magnetic tapes such as video tape, audio tape, and computer tape. The other markets for PET films are in packaging, electrical, metallized film, stationery and drafting, and tapes and labels. PET resins are also in commercial use in the form of blow-molded bottles. They are used as soft drink bottles in both the 32-ounce and 64-ounce sizes.

PET resins are made from approximately equal molar proportions of ethylene glycol and either dimethyl terephthalate (DMT) or terephthalic acid (TPA). This resin is essentially the same one that is used to produce polyester fiber. About 80% of PET film in the United States is based on DMT, and the remaining 20% is based on TPA.

Manufacturing Processes

1 PET RESIN FROM DMT AND ETHYLENE GLYCOL

Material Balance

Chemical	Coefficient T/T Product
Dimethyl terephthalate	−1.00
Ethylene glycol	−0.36
Methanol	0.34
PET resin	1.00

Primary energy requirements for utilities	0.22 FOET/T
Unit investment for a 45 kT plant (1977 $)	0.71 $1000/T

2 PET RESIN FROM TPA AND ETHYLENE GLYCOL

Material Balance

Chemical	Coefficient T/T Product
Ethylene glycol	−0.36
PET resin	1.00
Terephthalic acid (fiber grade)	−0.86

Primary energy requirements for utilities	0.26 FOET/T
Unit investment for a 45 kT plant (1977 $)	0.65 $1000/T

POLYPROPYLENE

Polypropylene (PP) is the fastest growing plastics material. It is a particularly useful resin because it can be made in a wide range of properties and be tailored to specific applications. Polypropylene resin is converted to end products primarily by injection molding and extrusion processes. Markets for polypropylene injection molding resin include appliances, consumer products, rigid packaging, and transportation. Extruded applications for polypropylene resin include film, and fibers and filaments which is the second largest market for this resin after injection molding products. Its major uses are in carpet backing, carpet face, ropes and cordage, upholstery and other fabrics, and strapping, bristles, and decorative ribbons.

Manufacturing Processes

1 PP BY LIQUID PHASE PROCESS (DART TECHNOLOGY)

Material Balance

Chemical	Coefficient T/T Product
Heptane	−0.03
Hydrogen chloride	−0.01
Isopropanol	−0.04
Polypropylene	1.00
Propylene (polymer grade)	−1.04
Sodium hydroxide	−0.01
Miscellaneous chemicals	−0.01

Primary energy requirements for utilities	0.31 FOET/T
Unit investment for a 91 kT plant (1977 $)	0.79 $1000/T

Comments Propylene is polymerized using titanium trichloride-aluminum alkyl catalyst.

2 PP BY VAPOR PHASE PROCESS (BASF TECHNOLOGY)

Material Balance

Chemical	Coefficient T/T Product
Cyclohexane	−0.03
Polypropylene	1.00
Propylene (polymer grade)	−1.02

Primary energy requirements for utilities	0.25 FOET/T
Unit investment for a 91 kT plant (1977 $)	0.66 $1000/T

Comments Gas phase polymerization is catalyzed by a modified Ziegler-Natta catalyst.

3 PP BY SLURRY PROCESS

Material Balance

Chemical	Coefficient T/T Product
Heptane	−0.02
Methanol	−0.02
Polypropylene	1.00
Propylene (polymer grade)	−1.12
Sodium hydroxide	−0.01
Miscellaneous chemicals	−0.01

Primary energy requirements for utilities	0.77 FOET/T
Unit investment for a 91 kT plant (1977 $)	0.83 $1000/T

4 PP BY SOLUTION PROCESS

Material Balance

Chemical	Coefficient T/T Product
Polypropylene	1.00
Propylene (polymer grade)	−1.12

Primary energy requirements for utilities	0.77 FOET/T
Unit investment for a 91 kT plant (1977 $)	0.87 $1000/T

5 LP BY NEW SLURRY PROCESS

Material Balance

Chemical	Coefficient T/T Product
Heptane	−0.02
Polypropylene	1.00
Propylene (polymer grade)	−1.08

Primary energy requirements for utilities	0.36 FOET/T
Unit investment for a 91 kT plant (1977 $)	0.64 $1000/T

POLYSTYRENE RESINS

The major market for polystyrene (PS) is in packaging which takes about one-third of its total outlet. The main consumers in this area are dairy containers, cups and lids, meat and poultry trays, egg cartons, fast food containers, bottles, and boxes, vials, jars, and tubes. Other than packaging, polystyrene resins are used in such products as toys, sporting goods, recreational articles, housewares, furnishings, consumer products, appliances and TV cabinets, disposable serviceware and flatware, construction, electrical and electronic parts, and furniture.

Generally, three different grades of polystyrene are produced in the United States: (1) General-purpose polystyrenes which are high-molecular-weight, glassy, and crystal clear resins. Injection molding is the most important fabrication process used for such resins, and typical end uses are in packaging, housewares, and a great variety of commercial items. (2) Impact polystyrenes which are rubber-modified resins and contain about 5% or less polybutadiene rubber. They are converted to end products mainly by injection molding and extrusion processes and used in products such as small appliance housings, toys, or packaging containers. (3) Expandable beads which are used for production of molded foam products.

Manufacturing Processes

1 CRYSTAL GRADE POLYSTYRENE BY BULK POLYMERIZATION

Material Balance

Chemical	Coefficient T/T Product
Polystyrene	1.00
Styrene	−1.02

Primary energy requirements for utilities   0.05 FOET/T
Unit investment for a 68 kT plant (1977 $)   0.30 $1000/T

Comments   Styrene is polymerized without a catalyst.

2 IMPACT GRADE POLYSTYRENE BY SUSPENSION POLYMERIZATION

Material Balance

Chemical	Coefficient T/T Product
Polybutadiene rubber	−0.05
Polystyrene	1.00
Styrene	−0.98
Miscellaneous chemicals	−0.03

Primary energy requirements for utilities   0.14 FOET/T
Unit investment for a 68 kT plant (1977 $)   0.47 $1000/T

3 IMPACT GRADE POLYSTYRENE BY BULK/SUSPENSION POLYMERIZATION

Material Balance

Chemical	Coefficient T/T Product
Polybutadiene rubber	−0.05
Polystyrene	1.00
Styrene	−0.97
Miscellaneous chemicals	−0.04

Primary energy requirements for utilities   0.21 FOET/T
Unit investment for a 68 kT plant (1977 $)   0.48 $1000/T

4 POLYSTYRENE EXPANDABLE BEADS BY SUSPENSION POLYMERIZATION

Material Balance

Chemical	Coefficient T/T Product
Pentane	−0.07
Polystyrene (crystal grade)	−0.05
Polystyrene (expandable beads)	1.00
Styrene	−0.93
Miscellaneous chemicals	−0.02

Primary energy requirements for utilities   0.11 FOET/T
Unit investment for an 18 kT plant (1977 $)   0.82 $1000/T

POLYURETHANE FOAMS

Polyurethane foams are among the most successful plastics materials produced in the United States. They are consumed as both flexible and rigid foams in a variety of different applications. Flexible foams account for more than two-thirds of polyurethanes' total outlet. The major markets for flexible polyurethane foams are in furniture, primarily in furniture cushioning; and transportation, mainly in passenger car seating, seating in other motor vehicles as well as boats and aircraft, and miscellaneous padding. Flexible foams are also used in bedding and rug underlay. Rigid polyurethane foams are mainly used in building insulation, and refrigerators and freezers. The other markets for rigid foams include furniture, industrial insulation, and transportation.

Polyurethane foams have been commercially produced since the mid-1950s from polyhydroxy components and polyisocyanates.

Manufacturing Processes

1 POLYURETHANE FLEXIBLE FOAM (SLABSTOCK)

Material Balance

Chemical	Coefficient T/T Product
Freon 11	−0.10
Polyether polyol-triol	−0.75
Polyurethane	1.00
Toluene diisocyanate	−0.33

Primary energy requirements for utilities	0.04 FOET/T
Unit investment for a 3 kT plant (1977 $)	1.43 $1000/T

2 POLYURETHANE FIRE RETARDANT RIGID FOAM (FOAM SHEET)

Material Balance

Chemical	Coefficient T/T Product
Freon 11	−0.15
Polyether polyol-hexol	−0.32
Polyether polyol-phosphorous	−0.12
Polyisocyanates (MDI/PMPPI)	−0.53
Polyurethane	1.00
Miscellaneous chemicals	−0.02

Primary energy requirements for utilities	0.04 FOET/T
Unit investment for a 3 kT plant (1977 $)	1.43 $1000/T

POLYVINYL ACETATE (PVAc) AND VINYL ACETATE COPOLYMERS

Polyvinyl acetate (PVAc) is primarily used in copolymer form; the major copolymers are those with n-butyl acrylate, and 2-ethylhexyl acrylate, and those with ethylene. Vinyl acetate homo- and copolymers are used mainly in the form of aqueous emulsions. However, some of them come as solid resins in the form of powder or beads.

Latex paints make up the largest single market for PVAc emulsions; practically all are copolymers. The second largest market for PVAc emulsions and resins is in adhesives which are mainly used for packaging and labeling, and construction. They are also used in areas such as paper coatings, textile treating, and nonwoven binders.

Manufacturing Processes

1 PVAc BY SOLUTION POLYMERIZATION

Material Balance

Chemical	Coefficient T/T Product
Methanol	−0.03
Polyvinyl acetate	1.00
Vinyl acetate	−1.02
Miscellaneous chemicals	−0.01

Primary energy requirements for utilities	0.43 FOET/T
Unit investment for a 45 kT plant (1977 $)	0.60 $1000/T

Comments   The reaction is conducted continuously in two-stage polymerizers, using solution polymerization with methanol as a solvent and benzoyl peroxide as an initiator.

2 PVAc BY SUSPENSION POLYMERIZATION

Material Balance

Chemical	Coefficient T/T Product
Acetaldehyde	−0.02
Hydrogen chloride	−0.01
Polyvinyl acetate	1.00
Vinyl acetate	−1.02
Miscellaneous chemicals	−0.02

Primary energy requirements for utilities   0.07 FOET/T

Unit investment for a 45 kT plant (1977 $)   0.40 $1000/T

Comments The polymerization is initiated with benzoyl peroxide and hydrogen peroxide, and the molecular weight is controlled with acetaldehyde as a chain transfer agent.

3 PVAc LATEX (62‰ SOLIDS) BY EMULSION POLYMERIZATION

Material Balance

Chemical	Coefficient T/T Product
Polyvinyl acetate (latex)	1.00
Vinyl acetate	−1.02
Miscellaneous chemicals	−0.12

Primary energy requirements for utilities   0.06 FOET/T

Unit investment for a 45 kT plant (1977 $)   0.31 $1000/T

Comments The reaction is initiated with potassium persulfate and continued in two stages. In the first stage, vinyl acetate conversion of 95% is obtained.

4 VINYL ACETATE/ETHYLENE COPOLYMER LATEX (60‰ SOLIDS) BY EMULSION POLYMERIZATION

Material Balance

Chemical	Coefficient T/T Product
Ethylene	−0.15
Polyvinyl alcohol	−0.05
Vinyl acetate	−0.87
Vinyl acetate/ethylene	1.00
Miscellaneous chemicals	−0.01

Primary energy requirements for utilities   0.06 FOET/T

Unit investment for a 45 kT plant (1977 $)   0.51 $1000/T

Comments The emulsion polymerization is conducted in two stages. The first stage operates at 140°F and 42 atm and the second stage operates at atmospheric pressure and the same temperature as first stage.

POLYVINYL ALCOHOL

Polyvinyl alcohols (PVA) are produced in several grades by hydrolysis of polyvinyl acetate. PVA's most important use in the United States is in textile wrap sizing and finishing. Adhesives is the second largest outlet for this material; it is used as an additive to both polyvinyl acetate and starch-based adhesives used in bag making, carton sealing, tube winding, and solid board lamination. Polyvinyl alcohol is also used in paper sizing and coatings where it provides good water barrier properties.

Manufacturing Processes

1 PVA FROM VINYL ACETATE

Material Balance

Chemical	Coefficient T/T Product
Acetic acid	1.31
Ethyl acetate	−0.07
Methanol	−0.01
Polyvinyl alcohol	1.00
Sodium hydroxide	−0.06
Sulfuric acid	−0.07
Vinyl acetate	−1.98
Miscellaneous chemicals	−0.01

Primary energy requirement for utilities	1.96 FOET/T
Unit investment for a 45 kT plant (1977 $)	1.48 $1000/T

POLYVINYL BUTYRAL

Polyvinyl butyrals (PVB) are produced in several grades by reaction of completely hydrolyzed polyvinyl acetate with butyraldehyde in the presence of a mineral acid. The largest single end use for polyvinyl butyral is as adhesive film in the lamination of auto safety glass. They are also used as a component of rather specialized textile finishes.

Manufacturing Processes

1 PVB BY CONDENSATION OF POLYVINYL ALCOHOL WITH BUTYRALDEHYDE

Material Balance

Chemical	Coefficient T/T Product
Butyraldehyde	−0.41
Ethyl acetate	−0.05
Methanol	−0.13
Polyvinyl butyral	1.00
Sodium hydroxide	−0.10
Sulfuric acid	−0.12
Vinyl acetate	−1.40

Primary energy requirements for utilities	3.00 FOET/T
Unit investment for an 11 kT plant (1977 $)	2.41 $1000/T

POLYVINYL CHLORIDE AND VINYL CHLORIDE COPOLYMERS

Polyvinyl chloride (PVC) is a brittle resin that easily decomposes with the application of heat, but through compounding with plasticizers, stabilizers, fillers, and other additives, it has become a major commercial resin. The majority of vinyl chloride resins are homopolymers, but copolymers are still essential in some applications. The most important commercial copolymers of vinyl chloride are those with vinyl acetate which are used in a variety of different applications such as phonograph records, calendering and extrusion formulations, and coating and adhesives.

PVC resins are converted to end products using different conversion processes such as rigid and flexible extrusion, rigid and flexible calendering, dispersion processes, injection molding, compression molding, and blow molding. Among the extruded products rigid pipe and tubing, wire and cable, and film and sheet for packaging are major products used in consumer market. Products of calendering processes include film and sheeting, flooring, and textile. The majority of flexible film and sheeting produced is combined with textiles to produce vinyl coated fabrics. Calendered rigid PVC sheet is used for credit cards, laminates with plywood or particleboard, and packaging. Products of dispersion processes have also found their way in areas such as plastisol formulating and molding, textile and paper coating, and flooring. Molded products are used in phonograph records, pipe fittings, automobile bumpers, toys, shoe soles and heels, bicycle grips, bottles, and a variety of other applications. PVC resins are also used in solution applications as protective coatings and adhesives.

Manufacturing Processes

1 PVC BY BULK POLYMERIZATION

Material Balance

Chemical	Coefficient T/T Product
PVC	1.00
Vinyl chloride	−1.03

Primary energy requirements for utilities   0.10 FOET/T

Unit investment for a 100 kT plant (1977 $)   0.55 $1000/T

2 PVC BY SUSPENSION POLYMERIZATION

Material Balance

Chemical	Coefficient T/T Product
PVC	1.00
Vinyl chloride	−1.03
Miscellaneous chemicals	−0.01

Primary energy requirements for utilities   0.31 FOET/T

Unit investment for a 100 kT plant (1977 $)   0.62 $1000/T

Comments The reaction is initiated by acetylcyclohexyl sulfonyl peroxide (ACSP) and diisopropyl peroxydicarbonate (DIPC) and carried out in batch reactors.

3 PVC (LATEX) BY EMULSION POLYMERIZATION

Material Balance

Chemical	Coefficient T/T Product
PVC (latex)	1.00
Vinyl chloride	−1.03
Miscellaneous chemicals	−0.03

Primary energy requirements for utilities   0.39 FOET/T

Unit investment for a 100 kT plant (1977 $)   0.79 $1000/T

Comments The polymerization, initiated with hydrogen peroxide and ascorbic acid, is continuously carried out and completed with 90% conversion.

4 PVC (LATEX) BY EMULSION POLYMERIZATION (BATCH PROCESS)

Material Balance

Chemical	Coefficient T/T Product
PVC (latex)	1.00
Vinyl chloride	−1.03
Miscellaneous chemicals	−0.03

Primary energy requirements for utilities	0.47 FOET/T
Unit investment for a 45 kT plant (1977 $)	0.97 $1000/T

5 VINYL CHLORIDE/VINYL ACETATE COPOLYMER BY SUSPENSION POLYMERIZATION

Material Balance

Chemical	Coefficient T/T Product
Vinyl acetate	−0.19
Vinyl chloride	−0.88
Vinyl chloride/vinyl acetate	1.00
Miscellaneous chemicals	−0.02

Primary energy requirements for utilities	0.31 FOET/T
Unit investment for a 23 kT plant (1977 $)	0.83 $1000/T

STYRENE-ACRYLONITRILE (SAN)

Styrene-acrylonitrile (SAN) copolymers generally contain 24–26% acrylonitrile, and are clear and transparent. Essentially all SAN is injection molded. Their principal uses are for drinking tumblers and other houseware items such as blender jars and covers, and some dishes and trays in major appliances. These copolymers are also used for instrument panel windows and glass-filled structural parts in automobiles, instrument lenses, battery jars, and medical instruments.

Manufacturing Processes

1 SAN BY BULK POLYMERIZATION

Material Balance

Chemical	Coefficient T/T Product
Acrylonitrile	−0.26
SAN	1.00
Styrene	−0.75

Primary energy requirements for utilities	0.19 FOET/T
Unit investment for a 14 kT plant (1977 $)	0.57 $1000/T

2 SAN BY CONTINUOUS EMULSION POLYMERIZATION

Material Balance

Chemical	Coefficient T/T Product
Acrylonitrile	−0.26
Hydrogen chloride (dilute)	−0.01
SAN	1.00
Sodium hydroxide	−0.01
Styrene	−0.74
Miscellaneous chemicals	−0.03

Primary energy requirements for utilities	0.11 FOET/T
Unit investment for a 14 kT plant (1977 $)	0.50 $1000/T

3 SAN BY BATCH EMULSION POLYMERIZATION

Material Balance

Chemical	Coefficient T/T Product
Acrylonitrile	−0.26
Hydrogen chloride (dilute)	−0.01
SAN	1.00
Sodium hydroxide	−0.01
Styrene	−0.74
Miscellaneous chemicals	−0.03

Primary energy requirements for utilities	0.10 FOET/T
Unit investment for a 14 kT plant (1977 $)	0.47 $1000/T

4 SAN BY SUSPENSION POLYMERIZATION

Material Balance

Chemical	Coefficient T/T Product
Acrylonitrile	−0.26
SAN	1.00
Styrene	−0.74
Miscellaneous chemicals	−0.03

Primary energy requirements for utilities	0.08 FOET/T
Unit investment for a 14 kT plant (1977 $)	0.35 $1000/T

UNSATURATED POLYESTER RESINS

Unsaturated polyester resins, generally, appear in three different types: general purpose resins, corrosion resistant resins, and fire retardant resins. About 80% of these resins are used for the production of glass fiber reinforced plastics (FRP), and the remaining is consumed in nonreinforced applications. Non-FRP uses are mainly in synthetic marble, auto repair putty, and cast furniture parts. Reinforced plastics are used in areas such as marine (e.g., boats and marine accessories), construction (e.g., sheets and panels, sanitary ware), transportation (e.g., passenger cars, trucks, trailers, buses), corrosion resistant products (e.g., tanks, pipes), consumer goods (e.g., appliance components, recreational articles, and toys), and components for electrical generation, transmission, and distribution (e.g., light poles, transmission hardware).

Unsaturated polyester resins consist of the resinous product of a polycondensation reaction between dicarboxylic acids and glycols where one of the acids is unsaturated. This polycondensate is dissolved in a vinylic monomer, usually styrene, resulting in a fluid-mixture product.

Manufacturing Processes

1 GENERAL PURPOSE RESIN BY FUSION PROCESS FROM PROPYLENE GLYCOL AND ANHYDRIDES

Material Balance

Chemical	Coefficient T/T Product
Maleic anhydride	−0.16
Phthalic anhydride	−0.25
Propylene glycol	−0.26
Styrene	−0.40
Unsaturated polyester	1.00

Primary energy requirements for utilities   0.04 FOET/T

Unit investment for a 19 kT plant (1977 $)   0.25 $1000/T

2 GENERAL PURPOSE RESIN BY BATCH SOLVENT PROCESS

Material Balance

Chemical	Coefficient T/T Product
Maleic anhydride	−0.16
Phthalic anhydride	−0.25
Propylene glycol	−0.26
Styrene	−0.40
Unsaturated polyester	1.00

Primary energy requirements for utilities   0.04 FOET/T

Unit investment for a 12 kT plant (1977 $)   0.36 $1000/T

3 GENERAL PURPOSE RESIN FROM PROPYLENE OXIDE AND ANHYDRIDES (BATCH PROCESS)

Material Balance

Chemical	Coefficient T/T Product
Ethylene glycol	−0.03
Maleic anhydride	−0.16
Phthalic anhydride	−0.24
Propylene oxide	−0.19
Styrene	−0.40
Unsaturated polyester	1.00

Primary energy requirements for utilities   0.02 FOET/T

Unit investment for a 26 kT plant (1977 $)   0.20 $1000/T

Comments   The reaction is carried out for about 3 hours at 390°F.

4 GENERAL PURPOSE RESIN FROM PROPYLENE OXIDE AND ANHYDRIDES (CONTINUOUS PROCESS)

Material Balance

Chemical	Coefficient T/T Product
Ethylene glycol	−0.03
Maleic anhydride	−0.16
Phthalic anhydride	−0.24
Propylene oxide	−0.19
Styrene	−0.40
Unsaturated polyester	1.00

Primary energy requirements for utilities   0.01 FOET/T
Unit investment for a 26 kT plant (1977 $)   0.19 $1000/T

5 GENERAL PURPOSE RESIN FROM PROPYLENE GLYCOL AND ANHYDRIDES (CONTINUOUS PROCESS)

Material Balance

Chemical	Coefficient T/T Product
Maleic anhydride	−0.16
Phthalic anhydride	−0.25
Propylene glycol	−0.26
Styrene	−0.40
Unsaturated polyester	1.00

Primary energy requirements for utilities   0.03 FOET/T
Unit investment for a 26 kT plant (1977 $)   0.20 $1000/T

6 CORROSION RESISTANT RESIN BY FUSION PROCESS FROM ISOPHTHALIC ACID

Material Balance

Chemical	Coefficient T/T Product
Isophthalic acid	−0.32
Maleic anhydride	−0.19
Propylene glycol	−0.31
Styrene	−0.30
Unsaturated polyester	1.00

Primary energy requirements for utilities   0.04 FOET/T
Unit investment for a 14 kT plant (1977 $)   0.21 $1000/T

7 FIRE RETARDANT RESIN BY FUSION PROCESS FROM PROPYLENE GLYCOL AND ANHYDRIDES

Material Balance

Chemical	Coefficient T/T Product
Maleic anhydride	−0.19
Phthalic anhydride	−0.13
Propylene glycol	−0.28
Styrene	−0.25
Tetrabromophthalic anhydride	−0.23
Unsaturated polyester	1.00

Primary energy requirements for utilities   0.04 FOET/T
Unit investment for a 14 kT plant (1977 $)   0.21 $1000/T

UREA-FORMALDEHYDE RESINS

Urea-formaldehyde resins account for about 80% of all the amino resins produced in the United States. These resins are used primarily for their adhesive properties in assembly gluing of furniture, bonding decorative laminates to plywood, and lumber core binding. They are also used in paper treating, surface coatings, textile treating and coating, and as molding powder in closures, and electrical appliance components.

Manufacturing Processes

1 UREA-FORMALDEHYDE MOLDING COMPOUND

Material Balance

Chemical	Coefficient T/T Product
Formaldehyde	−0.47
Sodium hydroxide	−0.01
Urea	−0.57
Urea-formaldehyde	1.00
Miscellaneous chemicals and additives	−0.46

Primary energy requirements for utilities   0.36 FOET/T
Unit investment for a 25 kT plant (1977 $)   1.12 $1000/T

2 UREA FORMALDEHYDE RESIN SYRUP (65‰ SOLIDS)

Material Balance

Chemical	Coefficient T/T Product
Formaldehyde	−0.47
Urea	−0.57
Urea-formaldehyde	1.00
Miscellaneous chemicals and additives	−0.01

Primary energy requirements for utilities   0.05 FOET/T
Unit investment for a 25 kT plant (1977 $)   0.18 $1000/T

Comments The polymerization is catalyzed by sodium hydroxide at pH 8 and carried out for 3–5 hours.

VINYLIDENE CHLORIDE COPOLYMERS

Vinylidene chloride has been copolymerized primarily with vinyl chloride, acrylic and methacrylic acid esters, and acrylonitrile. Because of their good characteristics, such as chemical and solvent resistance, flame retardance, impermeability to water vapor and gases, and toughness, these materials are used in such applications as packaging film, fibers, adhesives, cement additives, photographic films, magnetic tapes, tubing and gaskets, and pipe fitting, valves, and spinnerets.

Manufacturing Processes

1 VINYLIDENE CHLORIDE/VINYL CHLORIDE RESIN BY SUSPENSION POLYMERIZATION

Material Balance

Chemical	Coefficient T/T Product
Vinyl chloride	−0.17
Vinylidene chloride	−0.83
Vinylidene chloride/vinyl chloride	1.00
Miscellaneous chemicals	−0.01

Primary energy requirements for utilities	0.15 FOET/T
Unit investment for a 23 kT plant (1977 $)	1.06 $1000/T

2 VINYLIDENE CHLORIDE/ETHYL ACRYLATE/METHYL METHACRYLATE LATEX BY EMULSION POLYMERIZATION

Material Balance

Chemical	Coefficient T/T Product
Ethyl acrylate	−0.07
Methyl methacrylate	−0.03
Vinylidene chloride	−0.90
Terpolymer latex	1.00
Miscellaneous chemicals	−0.05

Primary energy requirements for utilities	0.09 FOET/T
Unit investment for a 23 kT plant (1977 $)	0.67 $1000/T

CHAPTER EIGHT

Man-Made Fibers

ACRYLIC AND MODACRYLIC FIBERS

Fibers containing 35% or more of the monomer acrylonitrile are commonly called acrylic fibers regardless of their acrylonitrile content. The U.S. Federal Trade Commission, however, classifies these fabrics into two groups, acrylic fibers, containing at least 85% by weight of acrylonitrile, and modacrylic fibers, containing less than 85% but at least 35% acrylonitrile. Modacrylic fibers have a significantly lower softening point than acrylic fibers; this property is probably the most important reason for their limited market penetration.

Acrylic and modacrylic fibers have become accepted because of their wool-like, soft feel properties which make them replacements for wool in a variety of household applications and apparel. Knit fabrics such as sweaters, pile fabrics, and hosiery consume about two thirds of the total production of these materials; home furnishings such as carpets and rugs, blankets, draperies and curtains, and upholstery use the remainder of acrylic and modacrylic fibers' outlet. These materials are used entirely in staple form and industrial uses are negligible.

Acrylic fibers contain typically 85-91% acrylonitrile, 7-8% of a neutral acrylate monomer such as methyl methacrylate or methyl acrylate, and 2-5% of other monomers and additives. Modacrylic fibers typically contain 37-81% acrylonitrile, and the rest are usually vinyl monomers such as vinyl chloride, and vinyl acetate. Solution and suspension polymerization are empolyed in the manufacture of these polymers.

Manufacturing Processes

1 ACRYLIC FIBER (POLYMER) BY CONTINUOUS SOLUTION POLYMERIZATION

Material Balance

Chemical	Coefficient T/T Product
Acrylic fiber	1.00
Acrylonitrile	−0.90
Methyl methacrylate	−0.08
Miscellaneous chemicals	−0.04

2 ACRYLIC FIBER (POLYMER) BY CONTINUOUS SOLUTION POLYMERIZATION

Material Balance

Chemical	Coefficient T/T Product
Acrylic fiber	1.00
Acrylonitrile	−0.90
Methyl acrylate	−0.08
Miscellaneous chemicals	−0.04

3 ACRYLIC FIBER (POLYMER) BY CONTINUOUS SUSPENSION POLYMERIZATION

Material Balance

Chemical	Coefficient T/T Product
Acrylic fiber	1.00
Acrylonitrile	−0.90
Methyl acrylate	−0.07
Miscellaneous chemicals	−0.04

4 ACRYLIC FIBER (POLYMER) BY BATCH SUSPENSION POLYMERIZATION

Material Balance

Chemical	Coefficient T/T Product
Acrylic fiber	1.00
Acrylonitrile	−0.90
Vinyl acetate	−0.09
Miscellaneous chemicals	−0.03

5 MODACRYLIC FIBER (POLYMER) BY CONTINUOUS SUSPENSION POLYMERIZATION

Material Balance

Chemical	Coefficient T/T Product
Acrylonitrile	−0.68
Modacrylic fiber	1.00
Vinyl chloride	−0.30
Miscellaneous chemicals	−0.02

6 MODACRYLIC FIBER (POLYMER) BY BATCH SOLUTION POLYMERIZATION

Material Balance

Chemical	Coefficient T/T Product
Acrylonitrile	−0.37
Isopropylacrylamide	−0.20
Methyl acrylate	−0.03
Modacrylic fiber	1.00
Vinylidene chloride	−0.40

7 MODACRYLIC FIBER (POLYMER) BY BATCH SUSPENSION POLYMERIZATION

Material Balance

Chemical	Coefficient T/T Product
Acrylonitrile	−0.77
Modacrylic fiber	1.00
Vinyl acetate	−0.20
Miscellaneous chemicals	−0.03

8 MODACRYLIC FIBER (POLYMER) BY BATCH SUSPENSION POLYMERIZATION

Material Balance

Chemical	Coefficient T/T Product
Acrylonitrile	−0.80
Modacrylic fiber	1.00
Vinyl bromide	−0.09
Vinylidene chloride	−0.08
Miscellaneous chemicals	−0.03

CELLULOSE ACETATE AND TRIACETATE FIBERS

Cellulose acetate is manufactured by acetylation of high-purity cellulose with a mixture of acetic acid and acetic anhydride. When acetate fibers were first introduced in 1924, they were thought of as artificial silk; they gained acceptance because of their appearance, pleasant touch, and excellent draping qualities. But after 1950, consumers switched to easy-care synthetic fibers such as polyester, acrylics, and polypropylene. Triacetate fibers which have an additional acetyl group and more closely resemble synthetic fibers, were introduced after 1954.

About 92% of all acetate and triacetate fibers are consumed in apparel. Knit fabrics such as dresses, blouses, and lingerie account for 68% of acetate and triacetate fibers consumed in apparel and the rest is used in woven fabrics. Home furnishings is the second largest end use for these fibers after apparel, taking about 14% of all the acetate and triacetate fibers' outlet. Draperies and curtains, comforters and bedspreads, and upholstery are the major consumption areas. Industrial and others consume the remainder of these fibers in areas such as fiberfill, paper and plastic tapes, electrical insulating tapes, and as colored electric-wire identification thread.

NYLON FIBERS

Nylon was the first commercially successful synthetic fiber introduced by DuPont in 1939. It was an immediate success in ladies' hosiery at the end of World War II and replaced silk and rayon in this application. Since then, nylon fibers have found their way in more important outlets such as lingerie, carpets and rugs, and stretch fabrics. They have also penetrated in large industrial markets, mainly in tire cord. Currently carpets and rugs are the largest market for these fibers. Over 60% of all the nylon fibers produced in the U.S. in 1977 were used in carpets and rugs. Apparel, particularly knit fabrics such as warp and weft fabrics, and women's and men's hosiery is the second largest market for nylon, taking about 17% of the total nylon fibers' outlet. Tire cord is the major industrial application for these fibers, accounting for 11% of their total consumption in 1977. The remainder of nylon fibers are used in other home furnishings and industrial applications such as upholstery, blankets, draperies and curtains, belting and hose, and seat belts.

Nylon is a polyamide, a class of compounds made by the condensation of polyfunctional organic amines and acids. The most important fiber types, however, are nylon 6, made from caprolactam, and nylon 66, the polymer of hexamethylenediamine and adipic acid.

POLYESTER FIBERS

Polyester fiber was originally developed in England by Calico Printers Association. DuPont made it a commercial fiber in 1953 under the trade name Dacron.

Polyester fiber has shown spectacular growth since 1960, mainly in blends with cotton for permanent-press fabrics used for men's shirts and slacks and ladies' blouses and sports clothes. Knit and woven fabrics, however, consumed about 65% of the polyester fiber's outlet in 1977. Home furnishings is the second largest end use for polyester fiber; carpets and rugs, and bed sheets and cases used almost 12% of this fiber's total production. Other home furnishings applications include draperies and curtains, blankets, and upholstery. Tire cord is the major industrial market for polyester fiber and it took 8% of the fiber's total outlet in 1977.

Polyester fiber is made from dimethyl terephthalate (DMT) or terephthalic acid (TPA) and ethylene glycol. Even though TPA-based fibers have several advantages such as less catalyst residue in the polymer, and higher-molecular-weight polymers (important in tire cord and other industrial fibers), they have a dye uniformity problem. As a result, essentially all polyester fibers derived from TPA are produced as staple or as industrial filament yarns.

POLYPROPYLENE FIBER

The largest end use for polypropylene fiber is in home furnishings which takes about 65% of the total polypropylene fiber's outlet. Carpets and rugs are responsible for about 80% of the fiber used in home furnishings. Industrial and other uses such as twine and cords, rope, bagging, outdoor furniture webs, and filtration, share almost equally the remainder of polypropylene fiber consumed in the United States.

RAYON

Rayon, a regenerated cellulose, was the first man-made fiber, developed in Europe in the late nineteenth century. Its commercial production in the United States began after World War I. Rayon was originally produced as a low-cost replacement for silk, but because of its high strength and good appearance it also competes with other natural fibers, mainly cotton. Rayon is increasingly replacing cotton in permanent-press blends, and it is expected to be used extensively in the fast-growing markets for nonwovens. In 1976, 21% of rayon's production was used in apparel, mainly broad woven fabrics (15%). Home furnishings accounted for 26% of its consumption mainly in draperies and curtains (14%), upholstery (7%) and mattress ticking (3%). Rayon was also used in industrial goods such as tire cord (6%) and hose (2%). However, competition from polyester and steel in tire cord may completely eliminate rayon in future years.

CHAPTER NINE

Synthetic Rubbers (Elastomers)

BUTYL RUBBER

Butyl rubber has been produced in the United States since 1942 by cationic copolymerization of isobutylene with isoprene (0.5–2.5 mole %) in the presence of Friedel-Crafts catalyst. Because of its outstanding air-retention properties, butyl rubber's major uses are in tires, tubes, and tire products. Applications in this area include use in inner tubes, innerliners for tubeless tires, tire bladders and valves, and tire sidewall components. Butyl rubber has also found its way in areas such as architectural and industrial sealants, sporting goods, nontire automotive goods, and liners for reservoirs and grain silos.

Manufacturing Processes

1 BUTYL RUBBER FROM ISOBUTYLENE

Material Balance

Chemical	Coefficient T/T Product
Butyl rubber	1.00
Isobutylene	−1.01
Isoprene	−0.03
Methyl chloride	−0.04

Primary energy requirements for utilities   1.27 FOET/T
Unit investment for a 41 kT plant (1977 $)   1.32 $1000/T

ETHYLENE-PROPYLENE RUBBERS

Ethylene and propylene can be copolymerized to give elastomers (EPR) whose outstanding properties are excellent abrasion, ozonolysis, and oxidation resistance. Neither ethylene nor propylene have more than a single double-bond. Therefore, EPR can only be cured by strong oxidizing agents such as peroxides. But because of several disadvantages, sulfur-curable EPR was introduced a decade ago by incorporating a third monomer containing two double-bonds. These elastomers are known as EPDM, for "ethylene-propylene-diene monomer" terpolymer. At the time, they were considered as another general purpose synthetic rubber. However, they have not yet been accepted in tire applications in any appreciable quantitites. In 1975, the tire market took over about 15% of all EPDM rubber; it is consumed for such uses as bicycle tires and sidewalls. Its nontire markets are primarily in molded automotive parts and other mechanical goods where it competes with neoprene (polychloroprene) and nitrile rubber.

Manufacturing Processes

1 EPR FROM ETHYLENE AND PROPYLENE

Material Balance

Chemical	Coefficient T/T Product
EPR	1.00
Ethylene	−1.41
Propylene (polymer grade)	−0.61
Miscellaneous chemicals	−0.03

Primary energy requirements for utilities	0.56 FOET/T
Unit investment for a 14 kT plant (1977 $)	1.47 $1000/T

2 EPDM BY SOLUTION POLYMERIZATION

Material Balance

Chemical	Coefficient T/T Product
EPDM	1.00
Ethylene	−0.44
n-Hexane	−0.01
Propylene (polymer grade)	−0.52
Sodium hydroxide	−0.01
Miscellaneous chemicals	−0.11

Primary energy requirements for utilities	1.07 FOET/T
Unit investment for a 27 kT plant (1977 $)	1.63 $1000/T

3 EPDM BY SUSPENSION POLYMERIZATION

Material Balance

Chemical	Coefficient T/T Product
EPDM	1.00
Ethylene	−0.52
Propylene (polymer grade)	−0.44
Sodium hydroxide	−0.01
Toluene	−0.01
Miscellaneous chemicals	−0.10

Primary energy requirements for utilities	0.77 FOET/T
Unit investment for a 27 kT plant (1977 $)	1.35 $1000/T

NEOPRENE (POLYCHLOROPRENE)

Neoprene is one of the first synthetic rubbers, produced successfully since 1931. It is known primarily for its high resilience and excellent resistance to ozone and weathering. Neoprene also possesses high strength and good resistance to abrasion, oxidants, oil, and aging. About 20% of the U.S. consumption is in automotive nontire products such as belts, hoses, and molded goods. General industrial and mechanical products consume 25%; wire and cable jackets, 10%; construction, 10%; and adhesives, sealants, and coating, 10%.

Manufacturing Processes

1 NEOPRENE BY FREE RADICAL INITIATED EMULSION POLYMERIZATION

Material Balance

Chemical	Coefficient T/T Product
Butadiene	−0.80
Chlorine	−1.00
Neoprene	1.00
Sodium hydroxide	−0.74

Primary energy requirements for utilities	0.80 FOET/T
Unit investment for a 34 kT plant (1977 $)	1.54 $1000/T

NITRILE RUBBER

Nitrile rubber was produced in the United States since 1939 by the free radical initiated emulsion copolymerization of butadiene with acrylonitrile. Its unique oil-resistance property makes nitrile rubber the preferred product in such applications as self-sealing fuel tanks, gasoline hose, gaskets, and printing rolls. Approximately 20% of all nitrile rubber produced is in latex form. About 50% of the latex is used in paper impregnation, 40% in leather and textile treatment, and 10% in adhesives.

Manufacturing Processes

1 NITRILE RUBBER BY COLD EMULSION POLYMERIZATION

Material Balance

Chemical	Coefficient T/T Product
Acrylonitrile	−0.12
Butadiene	−0.73
Nitrile rubber	1.00
Miscellaneous chemicals	−0.33

Primary energy requirements for utilities	0.20 FOET/T
Unit investment for a 68 kT plant (1977 $)	0.77 $1000/T

2 NITRILE RUBBER BY HOT EMULSION POLYMERIZATION

Material Balance

Chemical	Coefficient T/T Product
Acrylonitrile	−0.24
Butadiene	−0.56
Nitrile rubber (latex)	1.00
Miscellaneous chemicals	−0.11

Primary energy requirements for utilities	0.08 FOET/T
Unit investment for a 23 kT plant (1977 $)	0.80 $1000/T

POLYBUTADIENE

Commercial production of solution-polymerized polybutadiene rubbers (using Ziegler-Natta-type catalyst) in the United States began in 1961. About 90% of cis-polybutadiene is used in tire tread, because of its good abrasion resistance, high resiliency, and excellent high- and low-temperature properties. On average, polybutadiene is used at levels of 40% of total elastomer content in passenger tire tread blends and 50% in truck tires. This product is also used in manufacture of high-impact polystyrene. Small amounts of polybutadiene are used in the production of industrial products such as belting, hoses, seals, gaskets, sheet packing, vibration dampers, play balls, golf balls, footwear, wire insulation, and sponge.

Manufacturing Processes

1 POLYBUTADIENE BY COBALT-CATALYZED POLYMERIZATION

Material Balance

Chemical	Coefficient T/T Product
Benzene	−0.04
Butadiene	−1.02
Polybutadiene	1.00
Miscellaneous chemicals and catalyst	−0.07

Primary energy requirements for utilities	0.74 FOET/T
Unit investment for a 91 kT plant (1977 $)	1.13 $1000/T

Comments Polymerization takes place in a mixed benzene/butene solvent in the presence of a cobalt-aluminum alkyl chloride catalyst. 80% of butadiene is converted to a polybutadiene of about 95% cis-1,4 structure.

2 POLYBUTADIENE BY LITHIUM-CATALYZED POLYMERIZATION

Material Balance

Chemical	Coefficient T/T Product
Butadiene	−1.02
Hexane	−0.03
Polybutadiene	1.00
Miscellaneous chemicals and catalyst	−0.02

Primary energy requirements for utilities	0.67 FOET/T
Unit investment for a 91 kT plant (1977 $)	0.96 $1000/T

Comments Butadiene polymerizes in a hexane solvent using butyllithium as catalyst at 50°C. Butadiene conversion is about 97%.

3 POLYBUTADIENE BY IODINE-ZIEGLER-CATALYZED POLYMERIZATION

Material Balance

Chemical	Coefficient T/T Product
Butadiene	−1.04
Polybutadiene	1.00
Toluene	−0.05
Miscellaneous chemicals and catalyst	−0.01

Primary energy requirements for utilities   1.14 FOET/T

Unit investment for a 91 kT plant (1977 $)   0.96 $1000/T

4 POLYBUTADIENE BY NICKEL-CATALYZED POLYMERIZATION

Material Balance

Chemical	Coefficient T/T Product
Butadiene	−1.03
Polybutadiene	1.00
Toluene	−0.02
Miscellaneous chemicals and catalyst	−0.04

Primary energy requirements for utilities   0.97 FOET/T

Unit investment for a 91 kT plant (1977 $)   0.84 $1000/T

POLYISOPRENE

Cis-polyisoprene is chemically identical to natural rubber and competes with it in both tire and nontire applications. In 1975, tire applications such as passenger and truck carcasses and truck treads consumed about 55% of polyisoprene's output. Other end uses include mechanical goods—13%, footwear—10%, rubber bands and erasers—7%, belting and hoses—5%, sporting goods—4%, and sealants and caulking compounds—3%.

Polyisoprene is produced by solution polymerization using Ziegler or lithium-type catalysts. Solvents employed in the process include normal paraffins (pentane, heptane range), isoparaffins, and aromatics, for example, benzene.

Manufacturing Processes

1 POLYISOPRENE BY ZIEGLER-CATALYZED POLYMERIZATION

Material Balance

Chemical	Coefficient T/T Product
Hexane	−0.01
Isoprene	−1.01
Polyisoprene	1.00
Miscellaneous chemicals and catalyst	−0.07

Primary energy requirements for utilities   0.84 FOET/T

Unit investment for a 45 kT plant (1977 $)   0.86 $1000/T

2 POLYISOPRENE BY LITHIUM-CATALYZED POLYMERIZATION

Material Balance

Chemical	Coefficient T/T Product
Isopentane	−0.01
Isoprene	−1.01
Polyisoprene	1.00
Miscellaneous chemicals and catalyst	−0.06

Primary energy requirements for utilities   0.73 FOET/T

Unit investment for a 45 kT plant (1977 $)   0.78 $1000/T

3 POLYISOPRENE LATEX FROM POLYISOPRENE

Material Balance

Chemical	Coefficient T/T Product
Polyisoprene	−1.01
Polyisoprene (latex)	1.00
Miscellaneous chemicals (emulsifier)	−0.16

Primary energy requirements for utilities   0.16 FOET/T

Unit investment for a 9 kT plant (1977 $)   0.36 $1000/T

STYRENE-BUTADIENE RUBBER (SBR)

Styrene-butadiene rubber (SBR) includes the various copolymers of butadiene and styrene produced by emulsion polymerization. SBR has also been produced by solution polymerization since 1971. Its major use is in tires (65-70%) where it is used primarily in passenger car treads because of wear characteristics superior to natural rubber. SBR is also used in nontire industrial products such as automotive applications—5%, nonautomotive mechanical goods—15%, and latex applications—10%.

Manufacturing Processes

1 SBR BY COLD EMULSION POLYMERIZATION

Material Balance

Chemical	Coefficient T/T Product
Butadiene	−0.73
SBR	1.00
Sodium chloride	−0.20
Styrene	−0.22
Sulfuric acid	−0.02
Miscellaneous chemicals	−0.11

Primary energy requirements for utilities	0.20 FOET/T
Unit investment for a 68 kT plant (1977 $)	0.77 $1000/T

Comments Reaction conversion is 60-65%. The unconverted butadiene is recovered by vacuum and the unconverted styrene by steam-stripping.

2 SBR BY SOLUTION POLYMERIZATION

Material Balance

Chemical	Coefficient T/T Product
Butadiene	−0.76
Hexane	−0.04
SBR	1.00
Styrene	−0.25
Miscellaneous chemicals and catalyst	−0.02

Primary energy requirements for utilities	0.38 FOET/T
Unit investment for a 68 kT plant (1977 $)	0.84 $1000/T

Comments The polymerization of styrene and butadiene in hexane solvent is catalyzed by butyllithium. Approximately 98% conversion is obtained in 4 hours.

3 SBR BY HOT EMULSION POLYMERIZATION

Material Balance

Chemical	Coefficient T/T Product
Butadiene	−0.56
SBR (latex)	1.00
Styrene	−0.46
Miscellaneous chemicals	−0.08

Primary energy requirements for utilities	0.08 FOET/T
Unit investment for a 23 kT plant (1977 $)	0.80 $1000/T

CHAPTER TEN

Thermoplastic Elastomers

STYRENE BLOCK COPOLYMERS

Styrene block copolymers are used mainly in injection-molded footwear (57% in 1976). Other injection-molded and extruded products such as wire and cable, toys and housewares, and miscellaneous industrial applications consume about 24%; adhesives use the remaining 19% of the styrene block copolymers' outlet.

These copolymers are available with variations in the molecular arrangement of the styrene and diene building blocks. The styrene content of the commercial styrene block copolymers varies between 25% and 50%. Typically, anionic solution polymerization is used in production of these materials using butyllithium catalysts and cyclohexane as solvent.

Manufacturing Processes

1 STAR-BLOCK STYRENE-BUTADIENE

Material Balance

Chemical	Coefficient T/T Product
Butadiene	−0.69
Cyclohexane	−0.02
Styrene	−0.30
Styrene-butadiene copolymer	1.00
Miscellaneous chemicals and catalyst	−0.02

Primary energy requirements for utilities	0.60 FOET/T
Unit investment for a 18 kT plant (1977 $)	1.09 $1000/T

Comments Product is a 200,000 MW star-block copolymer, 30% (by weight) styrene and the remaining 70% butadiene. Polymerization takes place in cyclohexane solvent at 158°F, initiated by *n*-butyllithium, modified by tetrahydrofuran.

THERMOPLASTIC COPOLYESTER ELASTOMERS

Thermoplastic copolyester is a new class of high-performance polymer, which is both thermoplastic and elastomeric. The bulk of the consumption of this product is divided into roughly equal portions for wire and cable, injection molding, and high-pressure hose. In injection molding, this material is used for a variety of small mechanical parts where it replaces other plastics such as nylons or polyacetals.

Thermoplastic copolyester is a three-block polycondensate with a crystalline phase of polybutylene terephthalate and a soft, amorphous phase of polytetramethylene ether glycol (PTMEG).

Manufacturing Processes

1 THERMOPLASTIC COPOLYESTER FROM DMT AND PTMEG

Material Balance

Chemical	Coefficient T/T Product
1,4-Butanediol	−0.25
Copolyester-ethers	1.00
Dimethyl terephthalate	−0.60
Miscellaneous chemicals (PTMEG)	−0.38

Primary energy requirements for utilities	0.32 FOET/T
Unit investment for a 9 kT plant (1977 $)	1.11 $1000/T

THERMOPLASTIC OLEFIN ELASTOMERS

Thermoplastic olefin elastomers are a combination of EPDM elastomer and polypropylene. On the average, their polypropylene content is 35–40%. Automotive components consume about 60% of all thermoplastic olefin elastomers used in the United States, wire and cable extrusion accounts for approximately 17%, and the rest is consumed in areas such as footwear, sporting goods, and electrical components.

Manufacturing Processes

1 THERMOPLASTIC OLEFIN ELASTOMERS FROM BLEND OF POLYPROPYLENE AND EPDM RUBBER

Material Balance

Chemical	Coefficient T/T Product
EPDM Rubber	−0.81
Polypropylene	−0.20
Thermoplastic olefin elastomer	1.00
Miscellaneous chemicals	−0.01

Primary energy requirements for utilities	0.17 FOET/T
Unit investment for an 18 kT plant (1977 $)	0.80 $1000/T

THERMOPLASTIC POLYURETHANES

These materials are used in a variety of different applications such as fabric coatings—33%, adhesives and binders—15%, exterior automotive parts—15%, and the rest is used in hydraulic hoses, cable jacketing, tubing, film, and other small automotive parts.

These thermoplastics are produced either in solution or in bulk by reacting linear polyhydroxy compounds with nearly stoichiometric amounts of isocyanates.

Manufacturing Processes

1 POLYESTER URETHANE ELASTOMER

Material Balance

Chemical	Coefficient T/T Product
Adipic acid	−0.41
1,4-Butanediol	−0.71
Polyisocyanates (MDI/PMPPI)	−0.33
Polyester urethane elastomer	1.00

Primary energy requirements for utilities	0.30 FOET/T
Unit investment for a 5 kT plant (1977 $)	1.45 $1000/T

2 POLYETHER URETHANE ELASTOMER

Material Balance

Chemical	Coefficient T/T Product
1,4-Butanediol	−0.06
Polyisocyanates (MDI/PMPPI)	−0.33
Polyether urethane elastomer	1.00
Tetrahydrofuran	−0.76

Primary energy requirements for utilities	0.33 FOET/T
Unit investment for a 5 kT plant (1977 $)	1.66 $1000/T

PART THREE

Index of Chemicals in the Technology Catalog

PRIMARY FEEDSTOCKS AND INTERMEDIATE CHEMICALS

T/T	Feedstock for	In Process	T/T	Produced in	In Process
Acetaldehyde					
0.78	Acetic acid	2	1.00	Acetaldehyde	1
1.08	Acetic anhydride	1	1.00	Acetaldehyde	2
0.02	Polyvinyl acetate	2	1.00	Acetaldehyde	3
0.63	Terephthalic acid	4	0.01	Vinyl acetate	2
Acetic Acid					
1.25	Acetic anhydride	2	1.00	Acetic acid	1
0.17	1,4-Butanediol	2	1.00	Acetic acid	2
0.04	1,4-Butanediol	3	1.00	Acetic acid	3
0.03	Ethylene glycol	2	1.00	Acetic acid	4
0.01	Isooctanol	1	0.21	Acetic anhydride	1
0.06	Terephthalic acid	1	0.05	Acrylic acid	1
0.70	Vinyl acetate	1	1.31	Polyvinyl alcohol	1
0.72	Vinyl acetate	2	0.26	Terephthalic acid	3
0.76	Vinyl acetate	3	0.55	Terephthalic acid	4
Acetic Anhydride					
0.10	Acetal resins	2	1.00	Acetic anhydride	1
			1.00	Acetic anhydride	2

329

T/T	Feedstock for	In Process	T/T	Produced in	In Process
Acetone					
0.03	Acetal resins	2	1.00	Acetone	1
0.28	Bisphenol-A	1	1.00	Acetone	2
1.22	Methyl isobutyl ketone	1	2.11	Hydrogen peroxide	2
0.68	Methyl methacrylate	1	0.61	Phenol	1
Acetylene					
0.42	Acrylic acid	2	1.00	Acetylene	1
0.32	1,4-Butanediol	1	1.00	Acetylene	2
0.68	Chloroprene	2	1.00	Acetylene	3
0.32	Ethyl acrylate	3	1.00	Acetylene	4
1.09	Ethylene	9	1.00	Acetylene	5
0.32	Vinyl acetate	2			
0.43	Vinyl chloride	3			
Acrolein					
1.04	Allyl alcohol	2	1.00	Acrolein	1
Acrylamide					
0.96	Polyacrylamide	1	1.00	Acrylamide	1
			1.00	Acrylamide	2
			1.00	Acrylamide	3
Acrylic Acid					
0.77	Ethyl acrylate	1	1.00	Acrylic acid	1
0.88	Methyl acrylate	1	1.00	Acrylic acid	2
Acrylonitrile					
0.76	Acrylamide	1	1.00	Acrylonitrile	1
0.78	Acrylamide	2	1.00	Acrylonitrile	2
0.92	Acrylamide	3			
0.90	Acrylic and modacrylic fibers	1			
0.90	Acrylic and modacrylic fibers	2			
0.90	Acrylic and modacrylic fibers	3			
0.90	Acrylic and modacrylic fibers	4			
0.68	Acrylic and modacrylic fibers	5			
0.37	Acrylic and modacrylic fibers	6			
0.77	Acrylic and modacrylic fibers	7			

T/T	Feedstock for	In Process	T/T	Produced in	In Process

Acrylonitrile (Continued)

T/T	Feedstock for	In Process	T/T	Produced in	In Process
0.80	Acrylic and modacrylic fibers	8			
0.19	Acrylonitrile-butadiene-styrene	1			
0.19	Acrylonitrile-butadiene-styrene	2			
0.22	Acrylonitrile-butadiene-styrene	3			
0.96	Adiponitrile	2			
0.73	Barrier resins	1			
0.57	Barrier resins	2			
0.79	Barrier resins	3			
0.58	Ethyl acrylate	2			
0.99	Hexamethylenediamine	1			
0.12	Nitrile rubber	1			
0.24	Nitrile rubber	2			
0.26	Styrene-acrylonitrile	1			
0.26	Styrene-acrylonitrile	2			
0.26	Styrene-acrylonitrile	3			
0.26	Styrene-acrylonitrile	4			

Active Carbon

T/T	Feedstock for	In Process	T/T	Produced in	In Process
0.02	Acrylamide	1			
0.02	Acrylamide	2			
0.01	Isopropanol	1			
0.01	Isopropanol	2			
0.02	Nylon resins	3			
0.02	Polyether polyols	5			

Adipic Acid

T/T	Feedstock for	In Process	T/T	Produced in	In Process
1.37	Adiponitrile	1	1.00	Adipic acid	1
1.41	Hexamethylenediamine	2	1.00	Adipic acid	2
0.65	Nylon resins	3			
0.41	Thermoplastic polyurethanes	1			

Adiponitrile

T/T	Feedstock for	In Process	T/T	Produced in	In Process
			1.00	Adiponitrile	1
			1.00	Adiponitrile	2

Allyl Alcohol

T/T	Feedstock for	In Process	T/T	Produced in	In Process
0.75	Glycerine	3	1.00	Allyl alcohol	1
			1.00	Allyl alcohol	2

T/T	Feedstock for	In Process	T/T	Produced in	In Process
Allyl Chloride					
0.98	Epichlorohydrin	1	1.00	Allyl chloride	1
1.00	Glycerine	1			
Aluminum Chloride					
0.02	Polybutenes and polyisobutylenes	1			
0.01	Polybutenes and polyisobutylenes	2			
Ammonia					
0.10	Acetal resins	1	1.00	Ammonia	1
0.63	Acrylamide	3	1.00	Ammonia	2
0.43	Acrylonitrile	1	0.86	Melamine	1
0.48	Adiponitrile	1	0.88	Melamine	2
0.22	Aniline	2	0.85	Melamine	3
0.22	Aniline	3	0.84	Melamine	4
0.01	Butadiene	1	0.02	Synthesis gas	3
1.32	Caprolactam	1	0.03	Synthesis gas	6
0.92	Caprolactam	2			
1.48	Caprolactam	3			
0.68	Caprolactam	4			
0.93	Caprolactam	5			
0.49	Hexamethylenediamine	2			
0.06	Hexamethylenediamine	3			
0.75	Hydrogen cyanide	1			
0.28	Nitric acid	1			
0.28	Nitric acid	2			
0.01	Propylene oxide	1			
0.57	Urea	1			
0.57	Urea	2			
Ammonium Sulfate					
			4.25	Caprolactam	1
			2.60	Caprolactam	2
			4.40	Caprolactam	3
			1.85	Caprolactam	4
			2.96	Caprolactam	5
Aniline					
0.80	Methylene diphenylene diisocyanate (MDI)	1	1.00	Aniline	1
			1.00	Aniline	2
			1.00	Aniline	3

T/T	Feedstock for	In Process	T/T	Produced in	In Process

Benzene

T/T	Feedstock for	In Process	T/T	Produced in	In Process
0.02	Bisphenol-A	1	1.00	Benzene	1
0.04	1,4-Butanediol	3	1.00	Benzene	2
0.02	1,4-Butanediol	4	0.03	Styrene	1
0.77	Chlorobenzene	1			
0.78	Chlorobenzene	2			
0.67	Cumene	1			
0.94	Cyclohexane	1			
0.74	Ethylbenzene	1			
1.19	Maleic anhydride	1			
0.66	Nitrobenzene	1			
0.94	Phenol	4			
0.04	Polybutadiene	1			

Bisphenol-A

T/T	Feedstock for	In Process	T/T	Produced in	In Process
0.67	Epoxy resins	1	1.00	Bisphenol-A	1
0.74	Epoxy resins	2			
0.76	Epoxy resins	3			
0.91	Polycarbonate resins	1			
0.91	Polycarbonate resins	2			
0.91	Polycarbonate resins	3			
0.82	Polycarbonate resins	4			

Boric Acid

T/T	Feedstock for	In Process
0.01	Cyclohexanol	2
0.01	Cyclohexanol	3

Butadiene

T/T	Feedstock for	In Process	T/T	Produced in	In Process
0.25	Acrylonitrile-butadiene-styrene	1	1.00	Butadiene	1
			1.00	Butadiene	2
0.25	Acrylonitrile-butadiene-styrene	2	1.00	Butadiene	3
0.10	Barrier resins	1			
0.73	1,4-Butanediol	2			
0.71	Chloroprene	1			
0.63	Hexamethylenediamine	3			
0.80	Neoprene (polychloroprene)	1			
0.73	Nitrile rubber	1			
0.56	Nitrile rubber	2			
1.02	Polybutadiene	1			
1.02	Polybutadiene	2			
1.04	Polybutadiene	3			
1.03	Polybutadiene	4			
0.69	Styrene block copolymers	1			

T/T	Feedstock for	In Process	T/T	Produced in	In Process

Butadiene (Continued)

T/T	Feedstock for	In Process
0.73	Styrene-butadiene rubber	1
0.76	Styrene-butadiene rubber	2
0.56	Styrene-butadiene rubber	3

N-*Butane*

T/T	Feedstock for	In Process
0.83	Acetic acid	3
1.90	Butadiene	3
1.07	Isobutane	1
1.22	Maleic anhydride	2
0.50	Polybutenes	1
0.01	Polyethylene (L.D.)	3

1,2-*Butanediol*

T/T	Produced in	In Process
0.15	1,4-Butanediol	3

1,4-*Butanediol*

T/T	Feedstock for	In Process	T/T	Produced in	In Process
0.28	Polybutylene terephthalate	1	1.00	1,4-Butanediol	1
0.44	Polybutylene terephthalate	2	1.00	1,4-Butanediol	2
0.25	Thermoplastic copolyester		1.00	1,4-Butanediol	3
	elastomers	1	1.00	1,4-Butanediol	4
0.71	Thermoplastic polyurethanes	1			
0.06	Thermoplastic polyurethanes	2			

N-*Butanol*

T/T	Produced in	In Process
0.03	1,4-Butanediol	1
1.00	*n*-Butanol	1
1.00	*n*-Butanol	2
1.00	*n*-Butanol	3
0.05	*n*-Butyraldehyde	1
0.11	2-Ethylhexanol	1

S-*Butanol*

T/T	Feedstock for	In Process	T/T	Produced in	In Process
1.31	Allyl alcohol	2	1.00	*s*-Butanol	1
1.10	Methyl ethyl ketone	1			

T-*Butanol*

T/T	Produced in	In Process
0.09	Isoprene	2

Butene-1

T/T	Feedstock for	In Process
0.01	Polyethylene (H.D.)	1
0.02	Polyethylene (H.D.)	2

T/T	Feedstock for	In Process	T/T	Produced in	In Process
Butenes					
2.53	Isobutylene	1	0.09	Acetic acid	4
0.57	Polybutenes	1	0.06	Ethylene	1
			0.46	Ethylene	2
			0.37	Ethylene	3
			0.04	Ethylene	4
			0.08	Ethylene	5
			0.36	Ethylene	6
			0.53	Ethylene	7
			0.57	Ethylene	8
N-*Butylenes*					
1.01	Acetic acid	4	1.51	Isobutylene	1
1.46	Butadiene	1			
1.19	Butadiene	2			
1.14	s-Butanol	1			
0.84	Methyl ethyl ketone	2			
S-*Butyl Ether*					
			0.12	Methyl ethyl ketone	2
N-*Butyraldehyde*					
0.41	Polyvinyl butyral	1	1.00	*n*-Butyraldehyde	1
Calcium Carbonate					
4.46	Acetylene	2			
Calcium Chloride					
0.08	Allyl alcohol	2			
Calcium Oxide					
0.76	Epichlorohydrin	1			
1.47	Ethylene oxide	3			
0.78	Glycerine	1			
1.20	Propylene oxide	1			
Caprolactam					
0.95	Nylon resins	1	1.00	Caprolactam	1
0.95	Nylon resins	2	1.00	Caprolactam	2
			1.00	Caprolactam	3
			1.00	Caprolactam	4
			1.00	Caprolactam	5

T/T	Feedstock for	In Process	T/T	Produced in	In Process
Carbon Dioxide					
0.32	Methanol	2	3.64	Hydrogen	3
0.36	Methanol	3			
0.74	Urea	1			
0.74	Urea	2			
Carbon Monoxide					
0.61	Acetic acid	1	1.00	Carbon monoxide	1
0.51	Acrylic acid	2	1.00	Carbon monoxide	2
0.54	*n*-Butanol	1	7.84	Hydrogen	3
0.64	*n*-Butyraldehyde	1			
0.35	Ethyl acrylate	3			
0.63	2-Ethylhexanol	1			
0.27	Isooctanol	1			
0.33	Isooctanol	2			
0.30	Phosgene	1			
Chlorine					
1.32	Allyl chloride	1	1.00	Chlorine	1
0.79	Chlorobenzene	1			
0.89	Chloroprene	1			
0.90	Epichlorohydrin	1			
0.70	Ethylene dichloride	1			
1.88	Ethylene oxide	3			
0.93	Glycerine	1			
1.00	Neoprene (polychloroprene)	1			
0.73	Phosgene	1			
1.46	Propylene oxide	1			
0.61	Vinyl chloride	1			
1.03	Vinylidene chloride	1			
0.92	Vinylidene chloride	2			
3.01	Vinylidene chloride	3			
Chlorobenzene					
0.03	Methylene diphenylene diioscyanate (MDI)	1	1.00	Chlorobenzene	1
			1.00	Chlorobenzene	2
1.19	Phenol	2			
1.31	Phenol	3			
0.01	Toluene diisocyanate	1			
Chloroprene					
			1.00	Chloroprene	1
			1.00	Chloroprene	2

T/T	Feedstock for	In Process	T/T	Produced in	In Process

Coal

2.11	Synthesis gas	3			
2.48	Synthesis gas	6			

Coke

1.86	Acetylene	2			
0.01	Acetylene	4			

Cumene

1.35	Phenol	1	1.00	Cumene	1

Cyclohexane

0.01	Acetal resins	1	1.00	Cyclohexane	1
0.01	Acetal resins	2			
0.78	Adipic Acid	1			
1.06	Caprolactam	2			
0.91	Caprolactam	4			
1.64	Cyclohexanol	1			
1.00	Cyclohexanol	2			
0.91	Cyclohexanol	3			
1.05	Cyclohexanone	2			
0.03	Polypropylene	2			
0.02	Styrene block copolymers	1			

Cyclohexanol

0.01	Acetal resins	2	1.00	Cyclohexanol	1
0.74	Adipic acid	2	1.00	Cyclohexanol	2
1.25	Aniline	2	1.00	Cyclohexanol	3
1.09	Cyclohexanone	1			

Cyclohexanone

0.96	Caprolactam	5	0.38	Cyclohexanol	1
			0.07	Cyclohexanol	2
			1.00	Cyclohexanone	1
			1.00	Cyclohexanone	2

Cyclohexylamine

0.13	Aniline	2			

Dichlorobenzene

			0.11	Chlorobenzene	1

Dichloropropylenes

			0.27	Allyl chloride	1

T/T	Feedstock for	In Process	T/T	Produced in	In Process
Diethylene Glycol					
0.27	Polyether polyols	4	0.11	Ethylene glycol	1
Diisobutyl Carbinol					
0.02	Hydrogen peroxide	1			
Dimethyl Terephthalate					
0.56	Polybutylene terephthalate	1	1.00	Dimethyl terephthalate	1
0.89	Polybutylene terephthalate	2	1.00	Dimethyl terephthalate	2
1.00	Polyethylene terephthalate	1			
0.60	Thermoplastic copolyester elastomers	1			
Dinitrotoluene					
1.66	Toluene diamine	1	1.00	Dinitrotoluene	1
Dipropylene Glycol					
			0.10	Propylene glycol	1
Epichlorohydrin					
0.56	Epoxy resins	1	1.00	Epiclorohydrin	1
0.56	Epoxy resins	2			
0.39	Epoxy resins	3			
1.03	Glycerine	2			
Ethane					
3.00	Acetylene	5			
0.92	Ethylene	1			
1.30	Ethylene	4			
0.44	Vinyl acetate	3			
0.56	Vinylidene chloride	3			
Ethanol					
1.20	Acetaldehyde	3	1.00	Ethanol	1
0.05	Epoxy resins	2			
0.48	Ethyl acrylate	1			
0.53	Ethyl acrylate	2			
0.53	Ethyl acrylate	3			
1.75	Ethylene	10			
Ethyl Acetate					
0.01	Acetic anhydride	1			
0.01	Acrylic acid	1			

T/T	Feedstock for	In Process	T/T	Produced in	In Process
Ethyl Acetate (Continued)					
0.07	Polyvinyl alcohol	1			
0.05	Polyvinyl butyral	1			
Ethyl Acrylate					
0.64	Polyacrylates	1	1.00	Ethyl acrylate	1
0.16	Polyacrylates	2	1.00	Ethyl acrylate	2
0.07	Vinylidene chloride copolymers	2	1.00	Ethyl acrylate	3
Ethylbenzene					
1.15	Styrene	1	1.00	Ethylbenzene	1
1.14	Styrene	2			
Ethyl Chloride					
			0.08	Vinylidene chloride	3
Ethylene					
0.68	Acetaldehyde	1	1.15	Acetylene	1
0.68	Acetaldehyde	2	1.00	Ethylene	1
0.76	Acrylonitrile	2	1.00	Ethylene	2
0.75	Ethanol	1	1.00	Ethylene	3
0.27	Ethylbenzene	1	1.00	Ethylene	4
0.36	Ethylene dichloride	1	1.00	Ethylene	5
			1.00	Ethylene	6
0.32	Ethylene dichloride	2	1.00	Ethylene	7
0.51	Ethylene glycol	2	1.00	Ethylene	8
0.96	Ethylene oxide	1	1.00	Ethylene	9
0.88	Ethylene oxide	2	1.00	Ethylene	10
0.78	Ethylene oxide	3			
0.41	Ethylene-propylene rubbers	1			
0.44	Ethylene-propylene rubbers	2			
0.52	Ethylene-propylene rubbers	3			
1.02	Polyethylene (H.D.)	1			
1.02	Polyethylene (H.D.)	2			
1.02	Polyethylene (H.D.)	3			
1.04	Polyethylene (H.D.)	4			
1.02	Polyethylene (H.D.)	5			

T/T	Feedstock for	In Process	T/T	Produced in	In Process
Ethylene (Continued)					
1.02	Polyethylene (H.D.)	6			
1.03	Polyethylene (L.D.)	1			
1.03	Polyethylene (L.D.)	2			
1.03	Polyethylene (L.D.)	3			
0.15	Polyvinyl acetate and vinyl acetate copolymers	4			
0.39	Vinyl acetate	1			
0.48	Vinyl chloride	1			
0.09	Vinylidene chloride	1			
Ethylene Dichloride					
1.66	Vinyl chloride	2	1.00	Ethylene dichloride	1
0.83	Vinylidene chloride	1	1.00	Ethylene dichloride	2
			0.18	Ethylene oxide	3
Ethylene Glycol					
0.01	Isooctanol	2	1.00	Ethylene glycol	1
0.36	Polyethylene terephthalate	1	1.00	Ethylene glycol	2
0.36	Polyethylene terephthalate	2			
0.03	Unsaturated polyester resins	3			
0.03	Unsaturated polyester resins	4			
Ethylene Oxide					
0.08	Acetal resins	1	1.00	Ethylene oxide	1
0.87	Ethylene glycol	1	1.00	Ethylene oxide	2
0.75	Polyether polyols	4	1.00	Ethylene oxide	3
2-Ethylhexanol					
			1.00	2-Ethylhexanol	1
Formaldehyde					
0.75	1,4-Butanediol	1	1.00	Formaldehyde	1
0.62	Isoprene	2			
0.23	Melamine-formaldehyde resins	1			
0.33	Melamine-formaldehyde resins	2			

T/T	Feedstock for	In Process	T/T	Produced in	In Process

Formaldehyde (Continued)

T/T	Feedstock for	In Process	T/T	Produced in	In Process
0.16	Methylene diphenylene diisocyanate (MDI)	1			
0.25	Phenol-formaldehyde resins	1			
0.34	Phenol-formaldehyde resins	2			
0.47	Urea-formaldehyde resins	1			
0.47	Urea-formaldehyde resins	2			

Formic Acid

T/T	Feedstock for	In Process	T/T	Produced in	In Process
			0.06	Acetic Acid	4

Freon 11

T/T	Feedstock for	In Process	T/T	Produced in	In Process
0.10	Polyurethane foams	1			
0.15	Polyurethane foams	2			

Fuel Gas

T/T	Feedstock for	In Process	T/T	Produced in	In Process
			1.30	Acetylene	1
			0.84	Acetylene	4
			0.57	Acetylene	5
			0.01	Benzene	2
			0.10	n-Butanol	2
			0.05	n-Butanol	3
			0.03	Cyclohexane	1
			0.06	Ethanol	1
			0.48	Ethylene	1
			0.58	Ethylene	3
			0.18	Ethylene	4
			0.77	Ethylene	5
			0.50	Ethylene	6
			0.39	Ethylene	7
			1.24	Ethylene	9
			0.02	Isopropanol	1

Fuel Oil (High Sulfur)

T/T	Feedstock for	In Process	T/T	Produced in	In Process
8.34	Acetylene	1	0.01	Ethylbenzene	1
0.52	Synthesis gas	2			
0.73	Synthesis gas	5			
0.91	Synthesis gas	7			

T/T	Feedstock for	In Process	T/T	Produced in	In Process

Fuel Oil (Low Sulfur)

			0.18	Acetylene	5
			0.12	Acrolein	1
			0.38	n-Butanol	1
			0.03	n-Butanol	2
			0.03	n-Butanol	3
			0.33	s-Butanol	1
			0.05	Caprolactam	1
			0.07	Caprolactam	5
			1.09	Ethylene	2
			0.05	Ethylene	3
			0.25	Ethylene	6
			2.16	Ethylene	7
			1.25	Ethylene	8
			0.45	2-Ethylhexanol	1
			0.19	Isooctanol	1
			0.41	Isooctanol	2
			0.64	Isoprene	1
			0.23	Propylene oxide	2

Gas Oil

T/T	Feedstock for	In Process
4.50	Ethylene	2
6.02	Ethylene	7
5.53	Ethylene	8

Glycerine

T/T	Feedstock for	In Process	T/T	Produced in	In Process
0.21	Polyether polyols	2	1.00	Glycerine	1
0.03	Polyether polyols	3	1.00	Glycerine	2
			1.00	Glycerine	3

Heptane

T/T	Feedstock for	In Process
0.02	Polycarbonate resins	1
0.02	Polycarbonate resins	3
0.02	Polycarbonate resins	4
0.03	Polypropylene	1
0.02	Polypropylene	3
0.03	Polypropylene	5

Heptenes

T/T	Feedstock for	In Process
1.08	Isooctanol	1
1.39	Isooctanol	2

T/T	Feedstock for	In Process	T/T	Produced in	In Process
Hexamethylenediamine					
0.52	Nylon resins	3	1.00	Hexamethylenediamine	1
			1.00	Hexamethylenediamine	2
			1.00	Hexamethylenediamine	3
Hexamethylenetetramine					
0.10	Phenol-formaldehyde	1			
Hexane					
0.01	Chloroprene	2			
0.01	Ethylene-propylene rubber	2			
0.10	Polybutenes and polyisobutylenes	2			
0.03	Polyethylene (H.D.)	2			
0.03	Polyethylene (H.D.)	5			
0.04	Polyethylene (H.D.)	6			
0.03	Polybutadiene	2			
0.01	Polyisoprene	1			
0.04	Styrene-butadiene rubber	2			
Hydrogen					
0.01	Acrylamide	1	0.23	Carbon monoxide	1
0.07	Aniline	1	0.25	Carbon monoxide	2
0.06	Aniline	2	0.03	Chlorine	1
0.07	Benzene	1	0.02	Cyclohexanone	2
0.07	1,4-Butanediol	1	1.00	Hydrogen	1
0.04	1,4-Butanediol	2	1.00	Hydrogen	2
0.04	1,4-Butanediol	3	1.00	Hydrogen	3
0.03	1,4-Butanediol	4			
0.07	*n*-Butanol	1			
0.03	*n*-Butanol	3			
0.05	*n*-Butyraldehyde	1			
0.08	Caprolactam	1			
0.04	Caprolactam	2			
0.05	Caprolactam	3			
0.04	Caprolactam	5			
0.07	Cyclohexane	1			
0.31	Ethylene	9			
0.08	2-Ethylhexanol	1			
0.07	Hexamethylenediamine	1			
0.07	Hexamethylenediamine	2			

T/T	Feedstock for	In Process	T/T	Produced in	In Process

Hydrogen (Continued)

T/T	Feedstock for	In Process	T/T	Produced in	In Process
0.07	Hexamethylenediamine	3			
0.06	Hydrogen peroxide	1			
0.01	Isobutane	1			
0.04	Isooctanol	1			
0.07	Isooctanol	2			
0.11	Toluene diamine	1			
0.01	p-Xylene	2			

Hydrogen Chloride

T/T	Feedstock for	In Process	T/T	Produced in	In Process
0.02	Acetaldehyde	2	0.64	Allyl chloride	1
0.17	Acrylonitrile	2	0.46	Methylene diphenylene diisocyanate (MDI)	1
0.05	Caprolactam	4	0.48	Phenol	2
0.50	Chlorobenzene	2	0.61	Vinyl chloride	2
0.49	Chloroprene	2	0.43	Vinylidene chloride	1
0.04	Ethyl acrylate	3	0.47	Vinylidene chloride	2
0.08	Glycerine	1	2.11	Vinylidene chloride	3
0.28	Glycerine	2			
0.01	Polyacrylates	2			
0.18	Polycarbonate resins	1			
0.03	Polycarbonate resins	2			
0.18	Polycarbonate resins	3			
0.17	Polycarbonate resins	4			
0.01	Polypropylene	1			
0.01	Polyvinyl acetate	2			
0.03	Propylene oxide	1			
0.60	Vinyl chloride	3			

Hydrogen Chloride (Dilute)

T/T	Feedstock for	In Process	T/T	Produced in	In Process
0.08	Acetone	2	0.40	Chlorobenzene	1
0.94	Ethylene dichloride	2			
0.47	Phenol	3			
0.01	Styrene-acrylonitrile	2			
0.01	Styrene-acrylonitrile	3			

Hydrogen Cyanide

T/T	Feedstock for	In Process	T/T	Produced in	In Process
0.60	Acrylonitrile	2	1.00	Hydrogen cyanide	1
0.61	Hexamethylenediamine	3			
0.32	Methyl methacrylate	1			

Hydrogen Peroxide

T/T	Feedstock for	In Process	T/T	Produced in	In Process
0.45	Glycerine	3	1.00	Hydrogen peroxide	1
			1.00	Hydrogen peroxide	2

T/T	Feedstock for	In Process	T/T	Produced in	In Process
Isobutane					
2.61	Propylene oxide	2	1.00	Isobutane	1
Isobutanol					
			0.12	1,4-Butanediol	3
			0.14	*n*-Butanol	1
			0.11	*n*-Butanol	2
			0.08	*n*-Butanol	3
			0.02	*n*-Butyraldehyde	1
			0.16	2-Ethylhexanol	1
Isobutylene					
1.01	Butyl rubber	1	1.00	Isobutylene	1
0.94	Isoprene	2	2.15	Propylene oxide	2
1.12	Methyl methacrylate	2			
1.01	Polybutenes and polyisobutylenes	2			
Isobutyraldehyde					
			0.25	*n*-Butyraldehyde	1
Isooctanol					
			1.00	Isooctanol	1
			1.00	Isooctanol	2
Isopentane					
0.01	Polyisoprene	2			
Isophthalic Acid					
0.32	Unsaturated polyester resins	6	1.00	Isophthalic acid	1
Isoprene					
0.03	Butyl rubber	1	1.00	Isoprene	1
1.01	Polyisoprene	1	1.00	Isoprene	2
1.01	Polyisoprene	2	1.00	Isoprene	3
Isopropanol					
1.11	Acetone	1	1.00	Isopropanol	1
2.26	Hydrogen peroxide	2	1.00	Isopropanol	2
0.04	Polypropylene	1			

T/T	Feedstock for	In Process	T/T	Produced in	In Process
Methanol					
1.48	Acetal resins	1	1.00	Methanol	1
1.32	Acetal resins	2	1.00	Methanol	2
0.58	Acetic acid	1	1.00	Methanol	3
0.11	1,4-Butanediol	3	0.18	Polybutylene terephthalate	1
0.41	Dimethyl terephthalate	1	0.29	Polybutylene terephthalate	1
0.35	Dimethyl terephthalate	2			
1.18	Formaldehyde	1	0.34	Polyethylene terephthalate	2
0.39	Methyl acrylate	1			
0.37	Methyl methacrylate	1			
0.38	Methyl methacrylate	2			
0.01	Nylon resins	3			
0.02	Polypropylene	3			
0.03	Polyvinyl acetate	1			
0.01	Polyvinyl alcohol	1			
0.13	Polyvinyl butyral	1			
Methyl Acetate					
			0.04	Acetic acid	2
Methyl Acrylate					
0.08	Acrylic and modacrylic fibers	2	1.00	Methyl acrylate	1
0.07	Acrylic and modacrylic fibers	3			
0.03	Acrylic and modacrylic fibers	6			
0.23	Barrier resins	1			
Methyl Chloride					
0.04	Butyl rubber	1			
Methylene Chloride					
0.01	Polycarbonate resins	1			
0.01	Polycarbonate resins	2			
0.01	Polycarbonate resins	3			
0.01	Polycarbonate resins	4			

T/T	Feedstock for	In Process	T/T	Produced in	In Process

Methylene Diphenylene Diisocyanate (MDI) and its Polymeric Form (PMPPI)

T/T	Feedstock for	In Process	T/T	Produced in	In Process
0.53	Polyurethane foams	2	1.00	Methylene diphenylene diisocyanate (MDI)	1
0.33	Thermoplastic polyurethanes	1			
0.33	Thermoplastic polyurethanes	2			

Methyl Ethyl Ketone

T/T	Feedstock for	In Process	T/T	Produced in	In Process
0.01	Acrylic acid	2	1.24	Allyl alcohol	2
0.24	Terephthalic acid	3	1.00	Methyl ethyl ketone	1
			1.00	Methyl ethyl ketone	2

Methyl Isobutyl Ketone

T/T	Feedstock for	In Process	T/T	Produced in	In Process
0.05	Epoxy resins	1	1.00	Methyl isobutyl ketone	1
0.04	Epoxy resins	3			

Methyl Methacrylate

T/T	Feedstock for	In Process	T/T	Produced in	In Process
0.08	Acrylic and modacrylic fibers	1	1.00	Methyl methacrylate	1
			1.00	Methyl methacrylate	2
0.04	Barrier resins	2			
0.31	Polyacrylates	1			
0.90	Polyacrylates	2			
1.00	Polyacrylates	3			
1.00	Polyacrylates	4			
0.03	Vinylidene chloride copolymers	2			

Methylnaphthalene

T/T	Feedstock for	In Process	T/T	Produced in	In Process
0.02	Hydrogen peroxide	1			

2-Methyl-1,3-Propanediol

T/T	Feedstock for	In Process	T/T	Produced in	In Process
			0.15	1,4-Butanediol	3
			0.26	1,4-Butanediol	4

Naphtha

T/T	Feedstock for	In Process	T/T	Produced in	In Process
4.31	Acetylene	4			
1.02	Ammonia	2			
0.80	Carbon monoxide	2			
3.25	Ethylene	3			
3.92	Ethylene	6			

T/T	Feedstock for	In Process	T/T	Produced in	In Process

Naphtha (Continued)

2.63	Hydrogen	2			
4.17	Hydrogen	3			
0.41	Synthesis gas	4			

Naphthalene

| 1.02 | Phthalic anhydride | 2 | | | |

Nitric Acid (95%)

| 0.54 | Dinitrotoluene | 1 | 1.00 | Nitric acid | 1 |

Nitric Acid (60%)

0.90	Adipic acid	1	1.00	Nitric acid	2
0.73	Adipic acid	2			
0.18	Dinitrotoluene	1			
0.54	Nitrobenzene	1			

Nitrobenzene

| 1.34 | Aniline | 1 | 1.00 | Nitrobenzene | 1 |

Nitrogen

| 0.04 | Acetone | 1 | | | |
| 0.09 | Methyl ethyl ketone | 1 | | | |

Oleum

3.15	Caprolactam	1			
2.10	Caprolactam	2			
1.36	Caprolactam	3			
1.39	Caprolactam	4			
2.76	Caprolactam	5			
1.19	Phenol	4			

Oxygen

0.40	Acetaldehyde	1			
7.67	Acetylene	1			
4.80	Acetylene	3			
4.31	Acetylene	4			
0.46	Acrylic acid	1			
0.90	Acrylonitrile	2			
0.04	Aniline	3			
0.48	1,4-Butanediol	3			
0.55	Caprolactam	2			
0.49	Caprolactam	5			

T/T	Feedstock for	In Process	T/T	Produced in	In Process

Oxygen (Continued)

T/T	Feedstock for	In Process	T/T	Produced in	In Process
0.31	Ethylene glycol	2			
1.10	Ethylene oxide	2			
4.44	Hydrogen	3			
0.17	Nitric acid	1			
1.13	Propylene oxide	2			
0.53	Synthesis gas	2			
0.74	Synthesis gas	5			
0.93	Synthesis gas	7			
0.88	Terephthalic acid	3			
0.33	Vinyl acetate	1			

Pentane

T/T	Feedstock for	In Process	T/T	Produced in	In Process
0.03	Methyl methacrylate	2	0.02	Isobutane	1
0.05	Polyethylene (H.D.)	1			
0.07	Polystyrene resins	4			

Pentenes

T/T	Feedstock for	In Process	T/T	Produced in	In Process
1.00	Isoprene	3			

Phenol

T/T	Feedstock for	In Process	T/T	Produced in	In Process
1.04	Aniline	3	1.00	Phenol	1
0.89	Bisphenol-A	1	1.00	Phenol	2
0.92	Caprolactam	3	1.00	Phenol	3
0.90	Phenol-formaldehyde resins	1	1.00	Phenol	4
0.83	Phenol-formaldehyde resins	2			

Phosgene

T/T	Feedstock for	In Process	T/T	Produced in	In Process
0.84	Methylene diphenylene diisocyanate (MDI)	1	1.00	Phosgene	1
0.42	Polycarbonate resins	1			
0.44	Polycarbonate resins	2			
0.42	Polycarbonate resins	3			
0.40	Polycarbonate resins	4			
1.26	Toluene diisocyanate	1			

Phosphoric Acid

T/T	Feedstock for	In Process	T/T	Produced in	In Process
0.01	Hydrogen cyanide	1			
0.23	Polyether polyols	2			
0.01	Polyether polyols	4			

T/T	Feedstock for	In Process	T/T	Produced in	In Process

Phthalic Anhydride

T/T	Feedstock for	In Process	T/T	Produced in	In Process
0.25	Unsaturated polyester resins	1	1.00	Phthalic anhydride	1
			1.00	Phthalic anhydride	2
0.25	Unsaturated polyester resins	2			
0.24	Unsaturated polyester resins	3			
0.24	Unsaturated polyester resins	4			
0.25	Unsaturated polyester resins	5			
0.13	Unsaturated polyester resins	7			

Propane

T/T	Feedstock for	In Process	T/T	Produced in	In Process
0.92	Ethylene	1	0.33	Propylene	1
2.36	Ethylene	5	0.43	Propylene	2

Propylene (Chemical Grade)

T/T	Feedstock for	In Process	T/T	Produced in	In Process
0.85	Acetone	2	0.14	Ethylene	1
0.72	Acrylic acid	1	0.70	Ethylene	2
1.20	Acrylonitrile	1	0.63	Ethylene	3
0.73	Allyl chloride	1	0.04	Ethylene	4
0.76	1,4-Butanediol	3	0.24	Ethylene	5
0.92	*n*-Butanol	1	0.60	Ethylene	6
0.74	*n*-Butanol	2	0.85	Ethylene	7
0.69	*n*-Butanol	3	0.80	Ethylene	8
0.86	*n*-Butyraldehyde	1	1.00	Propylene	1
0.38	Cumene	1			
1.04	2-Ethylhexanol	1			
2.08	Isoprene	1			
0.73	Isopropanol	1			
0.83	Isopropanol	2			
0.83	Propylene oxide	1			
0.78	Propylene oxide	2			
0.33	Styrene	2			

Propylene (Polymer Grade)

T/T	Feedstock for	In Process	T/T	Produced in	In Process
1.16	Acrolein	1	1.00	Propylene	2
0.61	Ethylene-propylene rubbers	1			
0.52	Ethylene-propylene rubbers	2			

T/T	Feedstock for	In Process	T/T	Produced in	In Process

Propylene (Polymer Grade) (Continued)

T/T	Feedstock for	In Process	T/T	Produced in	In Process
0.44	Ethylene-propylene rubbers	3			
0.02	Polyethylene (H.D.)	5			
0.02	Polyethylene (H.D.)	6			
1.04	Polypropylene	1			
1.02	Polypropylene	2			
1.12	Polypropylene	3			
1.12	Polypropylene	4			
1.08	Polypropylene	5			

Propylene (Refinery Grade)

T/T	Feedstock for	In Process	T/T	Produced in	In Process
1.33	Propylene	1			
1.43	Propylene	2			

Propylene Dichloride

T/T	Feedstock for	In Process	T/T	Produced in	In Process
			0.10	Propylene oxide	1

Propylene Glycol

T/T	Feedstock for	In Process	T/T	Produced in	In Process
0.06	Polyether polyols	5	1.00	Propylene glycol	1
0.26	Unsaturated polyester resins	1			
0.26	Unsaturated polyester resins	2			
0.26	Unsaturated polyester resins	5			
0.31	Unsaturated polyester resins	6			
0.28	Unsaturated polyester resins	7			

Propylene Oxide

T/T	Feedstock for	In Process	T/T	Produced in	In Process
1.11	Allyl alcohol	1	1.00	Propylene oxide	1
0.93	1,4-Butanediol	4	1.00	Propylene oxide	2
0.78	Polyether polyols	1	0.41	Styrene	2
0.65	Polyether polyols	2			
0.98	Polyether polyols	3			
0.98	Polyether polyols	5			
0.89	Propylene glycol	1			
0.19	Unsaturated polyester resins	3			
0.19	Unsaturated polyester resins	4			

T/T	Feedstock for	In Process	T/T	Produced in	In Process
Pyrolysis Gasoline					
			0.16	Ethylene	1
			0.74	Ethylene	2
			0.62	Ethylene	3
			0.05	Ethylene	4
			0.27	Ethylene	5
			1.21	Ethylene	6
			1.10	Ethylene	7
			1.40	Ethylene	8
Sodium Bicarbonate					
0.04	Isooctanol	2			
Sodium Bisulfate					
0.12	Adiponitrile	1			
0.12	Hexamethylenediamine	2			
Sodium Carbonate					
0.04	Chlorine	1			
0.04	Ethyl acrylate	3			
0.07	Glycerine	1			
0.07	Glycerine	2			
0.33	Polyacrylamide	1			
Sodium Chloride					
1.68	Chlorine	1			
0.20	Styrene-butadiene rubber	1			
Sodium Hydroxide					
0.06	Acrylamide	2	1.12	Chlorine	1
0.01	Allyl chloride	1			
0.01	Bisphenol-A	1			
0.30	Caprolactam	1			
0.01	Chlorobenzene	2			
0.66	Chloroprene	1			
0.13	Cyclohexanol	1			
0.35	Cyclohexanol	2			
0.07	Ethyl acrylate	1			
0.01	Ethylene	4			
0.04	Ethylene oxide	3			
0.01	Ethylene-propylene rubber	2			

T/T	Feedstock for	In Process	T/T	Produced in	In Process

Sodium Hydroxide (Continued)

T/T	Feedstock for	In Process
0.01	Ethylene-propylene rubber	3
0.24	Epoxy resins	1
0.25	Epoxy resins	2
0.20	Epoxy resins	3
0.49	Glycerine	1
0.49	Glycerine	2
0.01	Glycerine	3
0.01	Isoprene	2
0.07	Methyl acrylate	1
0.18	Methylene-diphenylene diisocyanate (MDI)	1
0.74	Neoprene	1
	(polychloroprene)	1
0.01	Phenol	1
1.06	Phenol	3
1.64	Phenol	4
0.03	Phenol-formaldehyde	2
0.09	Polybutenes and polyisobutylenes	1
0.01	Polybutenes and polyisobutylenes	2
0.54	Polycarbonate resins	1
0.39	Polycarbonate resins	2
0.54	Polycarbonate resins	3
0.51	Polycarbonate resins	4
0.02	Polyether polyols	4
0.01	Polypropylene	1
0.01	Polypropylene	3
0.06	Polyvinyl alcohol	1
0.10	Polyvinyl butyral	1
0.01	Styrene	2
0.01	Styrene-acrylonitrile	2
0.01	Styrene-acrylonitrile	3
0.01	Urea-formaldehyde	1
0.01	Vinyl chloride	1
0.01	Vinyl chloride	3
0.46	Vinylidene chloride	1

Sodium Hydroxide (Dilute)

T/T	Feedstock for	In Process
0.01	Acetone	2
0.02	s-Butanol	1

T/T	Feedstock for	In Process	T/T	Produced in	In Process
Sodium Hydroxide (Dilute) (Continued)					
0.10	Ethyl acrylate	3			
0.01	Isooctanol	1			
Sodium Sulfite					
			1.93	Phenol	4
Sorbitol					
0.23	Polyether polyols	1			
Styrene					
0.54	Acrylonitrile-butadiene-styrene	1	1.00	Styrene	1
			1.00	Styrene	2
0.54	Acrylonitrile-butadiene-styrene	2			
0.67	Acrylonitrile-butadiene-styrene	3			
0.23	Barrier resins	2			
0.26	Barrier resins	3			
1.02	Polystyrene resins	1			
0.98	Polystyrene resins	2			
0.97	Polystyrene resins	3			
0.93	Polystyrene resins	4			
0.75	Styrene-acrylonitrile	1			
0.74	Styrene-acrylonitrile	2			
0.74	Styrene-acrylonitrile	3			
0.74	Styrene-acrylonitrile	4			
0.30	Styrene block copolymers	1			
0.22	Styrene-butadiene rubber	1			
0.25	Styrene-butadiene rubber	2			
0.46	Styrene-butadiene rubber	3			
0.40	Unsaturated polyester resins	1			
0.40	Unsaturated polyester resins	2			
0.40	Unsaturated polyester resins	3			

T/T	Feedstock for	In Pro-cess	T/T	Produced in	In Pro-cess
Styrene (Continued)					
0.40	Unsaturated polyester resins	4			
0.40	Unsaturated polyester resins	5			
0.30	Unsaturated polyester resins	6			
0.25	Unsaturated polyester resins	7			
Succinic Acid					
			0.05	Adipic acid	1
Sulfur					
0.67	Caprolactam	3	0.09	Synthesis gas	3
0.33	Sulfuric acid	1	0.11	Synthesis gas	6
Sulfuric Acid					
0.04	Acetal resins	1	1.00	Sulfuric acid	1
0.07	Acrylamide	2			
1.77	Acrylamide	3			
0.15	Acrylonitrile	1			
0.01	Butadiene	1			
0.01	s-Butanol	1			
0.93	Caprolactam	5			
0.08	Chlorine	1			
0.03	Dinitrotoluene	1			
0.06	Ethyl acrylate	1			
1.20	Ethyl acrylate	2			
0.09	Hydrogen	3			
0.04	Isooctanol	2			
0.01	Isoprene	2			
0.02	Methanol	3			
0.06	Methyl acrylate	1			
1.63	Methyl methacrylate	1			
0.01	Methyl methacrylate	2			
0.03	Nitrobenzene	1			
0.01	Phenol	1			
0.04	Phenol-formaldehyde	2			
0.01	Polyether polyols	5			
0.07	Polyvinyl alcohol	1			
0.12	Polyvinyl butyral	1			
0.02	Styrene-butadiene rubber	1			

T/T	Feedstock for	In Process	T/T	Produced in	In Process
Synthesis Gas (H_2:CO = 1:1)					
0.49	1,4-Butanediol	3	1.00	Synthesis gas	1
0.44	1,4-Butanediol	4	1.00	Synthesis gas	2
0.51	*n*-Butanol	3			
Synthesis Gas (H_2:CO = 2:1)					
0.62	*n*-Butanol	2	5.47	Acetylene	1
			4.01	Acetylene	3
			1.00	Synthesis gas	3
			1.00	Synthesis gas	4
			1.00	Synthesis gas	5
Synthesis Gas (H_2:CO = 3:1)					
0.92	Methanol	2	1.00	Synthesis gas	6
0.89	Methanol	3	1.00	Synthesis gas	7
			1.00	Synthesis gas	8
Terephthalic Acid (Crude)					
0.88	Dimethyl terephthalate	2	1.00	Terephthalic acid	3
1.02	Terephthalic acid	2	1.00	Terephthalic acid	4
Terephthalic Acid (Fiber Grade)					
0.86	Polyethylene terephthalate	2	1.00	Terephthalic acid	1
			1.00	Terephthalic acid	2
Tetrabromophthalic Anhydride					
0.23	Unsaturated polyester resins	7			
Tetrahydrofuran					
0.76	Thermoplastic polyurethanes	2			
Toluene					
1.20	Benzene	1	0.05	Styrene	1
2.69	Benzene	2			
1.11	Caprolactam	1			
0.53	Dinitrotoluene	1			
0.01	Glycerine	1			
0.01	Glycerine	2			
0.01	Ethylene-propylene rubber	3			

T/T	Feedstock for	In Process	T/T	Produced in	In Process

Toluene (Continued)

0.05	Polybutadiene	3			
0.02	Polybutadiene	4			

Toluene Diamine

0.76	Toluene diisocyanate	1	1.00	Toluene diamine	1

Toluene Diisocyanate

0.33	Polyurethane foams	1	1.00	Toluene diisocyanate	1

Trichloroethylene

			0.13	Vinylidene chloride	2

Triethylene Glycol

			0.03	Ethylene glycol	1

Urea

3.04	Melamine	1	1.00	Urea	1
3.10	Melamine	2	1.00	Urea	2
3.13	Melamine	3			
3.04	Melamine	4			
0.57	Urea-Formaldehyde	1			
0.57	Urea-Formaldehyde	2			

Vinyl Acetate

0.09	Acrylic and modacrylic fibers	4	1.00	Vinyl acetate	1
			1.00	Vinyl acetate	2
0.20	Acrylic and modacrylic fibers	7	1.00	Vinyl acetate	3
1.02	Polyvinyl acetate	1			
1.02	Polyvinyl acetate	2			
1.02	Polyvinyl acetate	3			
0.87	Polyvinyl acetate	4			
1.98	Polyvinyl alcohol	1			
1.40	Polyvinyl butyral	1			
0.19	PVC and vinyl chloride copolymers	5			

Vinyl Bromide

0.09	Acrylic and modacrylic fibers	8			

T/T	Feedstock for	In Process	T/T	Produced in	In Process
Vinyl Chloride					
0.30	Acrylic and modacrylic fibers	5	1.00	Vinyl chloride	1
1.03	Polyvinyl chloride	1	1.00	Vinyl chloride	2
			1.00	Vinyl chloride	3
1.03	Polyvinyl chloride	2			
1.03	Polyvinyl chloride	3			
1.03	Polyvinyl chloride	4			
0.88	PVC and vinyl chloride copolymers	5			
0.72	Vinylidene chloride	2			
0.17	Vinylidene chloride copolymers	1			
Vinylidene Chloride					
0.40	Acrylic and modacrylic fibers	6	1.00	Vinylidene chloride	1
			1.00	Vinylidene chloride	2
0.08	Acrylic and modacrylic fibers	8	1.00	Vinylidene chloride	3
0.83	Vinylidene chloride copolymers	1			
0.90	Vinylidene chloride copolymers	2			
m-*Xylene*					
0.71	Isophthalic acid	1			
o-*Xylene*					
0.97	Phthalic anhydride	1			
p-*Xylene*					
0.63	Dimethyl terephthalate	1	1.00	p-Xylene	1
0.67	Terephthalic acid	1	1.00	p-Xylene	2
0.68	Terephthalic acid	3			
0.72	Terephthalic acid	4			
Xylenes					
1.15	p-Xylene	1	1.61	Benzene	2
1.25	p-Xylene	2			

END PRODUCTS

T/T	Feedstock for	In Process	T/T	Produced in	In Process
	Acetal Resins				
			1.00	Acetal resins	1
			1.00	Acetal resins	2
	Acrylic and Modacrylic Fibers				
			1.00	Acrylic fiber	1
			1.00	Acrylic fiber	2
			1.00	Acrylic fiber	3
			1.00	Acrylic fiber	4
			1.00	Modacrylic fiber	5
			1.00	Modacrylic fiber	6
			1.00	Modacrylic fiber	7
			1.00	Modacrylic fiber	8
	Acrylonitrile-Butadiene-Styrene (ABS)				
			1.00	ABS resin	1
			1.00	ABS resin	2
			1.00	ABS resin	3
	Barrier Resins				
			1.00	Barrier resin (Barex)	1
			1.00	Barrier resin (Cycopac)	2
			1.00	Barrier resin (Lopac)	3
	Butyl Rubber				
			1.00	Butyl rubber	1
	Copolyester-Ethers				
			1.00	Thermoplastic copolyester elastomers	1
	Epoxy Resins				
			1.00	Epoxy resins (liquid DGEBA)	1
			1.00	Epoxy resins (liquid DGEBA)	2
			1.00	Epoxy resins (solid DGEBA)	3
	Ethylene-Propylene Rubbers				
0.81	Thermoplastic olefin elastomers		1.00	EPDM rubber	2
		1	1.00	EPDM rubber	3
			1.00	EP rubber	1

T/T	Feedstock for	In Process	T/T	Produced in	In Process
	Melamine-Formaldehyde Resins				
			1.00	Melamine-formaldehyde (molding compounds)	1
			1.00	Melamine-formaldehyde (syrup)	2
	Neoprene (Polychloroprene)				
			1.00	Neoprene	1
	Nitrile Rubber				
			1.00	Nitrile rubber	1
			1.00	Nitrile rubber (latex)	2
	Nylon Resins				
			1.00	Nylon 6 (chips)	1
			1.00	Nylon 6 (melt)	2
			1.00	Nylon 66 (chips)	3
	Phenol-Formaldehyde Resins				
			1.00	Phenol-formaldehyde (molding compounds)	1
			1.00	Phenol-formaldehyde (syrup)	2
	Polyacrylamide				
			1.00	Polyacrylamide	1
	Polyacrylates				
			1.00	Polyacrylate (latex)	1
			1.00	Polyacrylate (pellets)	2
	Polybutadiene				
0.07	ABS resins	3	1.00	Polybutadiene	1
0.05	Polystyrene (impact grade)	2			
			1.00	Polybutadiene	2
0.05	Polystyrene (impact grade)	3	1.00	Polybutadiene	3
			1.00	Polybutadiene	4
	Polybutenes and Polyisobutylenes				
			1.00	Polybutenes	1
			1.00	Polyisobutylenes	2

T/T	Feedstock for	In Process T/T	Produced in	In Process
Polybutylene Terephthalate (PBT) Resins				
		1.00	PBT resin (glass-filled)	1
		1.00	PBT resin (plain)	2
Polycarbonate Resins				
		1.00	Polycarbonate resin	1
		1.00	Polycarbonate resin	2
		1.00	Polycarbonate resin	3
		1.00	Polycarbonate resin (flame-resistant)	4
Polyether Polyols				
0.32	Polyurethane (rigid foam)	2 1.00	Polyether polyol (Sorbitol-based hexol)	1
0.12	Polyurethane (rigid foam)	2 1.00	Polyether polyol (phosphorus containing)	2
0.75	Polyurethane (flexible foam)	1 1.00	Polyether polyol (glycerine-based triol)	3
Polyethylene (High Density)				
		1.00	Polyethylene (high-density)	1
		1.00	Polyethylene (high-density)	2
		1.00	Polyethylene (high-density)	3
		1.00	Polyethylene (high-density)	4
		1.00	Polyethylene (high-density)	5
		1.00	Polyethylene (high-density)	6
Polyethylene (Low Density)				
		1.00	Polyethylene (low-density)	1
		1.00	Polyethylene (low-density)	2
		1.00	Polyethylene (low-density)	3
Polyethylene Glycol				
		1.00	Polyether polyols	4

T/T	Feedstock for	In Process	T/T	Produced in	In Process
Polyethylene Terephthalate (PET) Resins					
			1.00	PET resin	1
			1.00	PET resin	2
Polyisoprene					
1.01	Polyisoprene (latex)	3	1.00	Polyisoprene	1
			1.00	Polyisoprene	2
			1.00	Polyisoprene (latex)	3
Polymethyl Methacrylate					
			1.00	Polyacrylates	3
			1.00	Polyacrylates	4
Polypropylene					
0.20	Thermoplastic olefin		1.00	Polypropylene	1
	elastomers	1	1.00	Polypropylene	2
			1.00	Polypropylene	3
			1.00	Polypropylene	4
			1.00	Polypropylene	5
Polypropylene Glycol					
			1.00	Polyether Polyols	5
Polystyrene Resins					
0.05	Polystyrene (expandable beads)	4	1.00	Polystyrene (crystal grade)	1
			1.00	Polystyrene (impact grade)	2
			1.00	Polystyrene (impact grade)	3
			1.00	Polystyrene (expandable beads)	4
Polyurethane Foams					
			1.00	Polyurethane (flexible foam)	1
			1.00	Polyurethane (rigid foam)	2
Polyvinyl Acetate (PVAc)					
			1.00	Polyvinyl acetate	1
			1.00	Polyvinyl acetate	2
			1.00	Polyvinyl acetate (latex)	3

T/T	Feedstock for	In Process	T/T	Produced in	In Process
Polyvinyl Alcohol					
0.05	Vinyl acetate/ ethylene copolymer (latex)	4	1.00	Polyvinyl alcohol	1
Polyvinyl Butyral					
			1.00	Polyvinyl butyral	1
Polyvinyl Chloride (PVC)					
			1.00	Polyvinyl chloride	1
			1.00	Polyvinyl chloride	2
			1.00	Polyvinyl chloride (latex)	3
			1.00	Polyvinyl chloride (latex)	4
Styrene-Acrylonitrile (SAN) Resins					
			1.00	SAN resin	1
			1.00	SAN resin	2
			1.00	SAN resin	3
			1.00	SAN resin	4
Styrene Block Copolymers					
			1.00	Styrene block copolymer	1
Styrene-Butadiene Rubber (SBR)					
0.21	Barrier resin (Cycopac)	2	1.00	SBR	1
			1.00	SBR	2
			1.00	SBR (latex)	3
Thermoplastic Olefin Elastomers					
			1.00	Thermoplastic olefin elastomers	1
Thermoplastic Polyurethanes					
			1.00	Polyester urethane elastomer	1
			1.00	Polyether urethane elastomer	2

T/T	Feedstock for	In Process T/T	Produced in	In Process
	Unsaturated Polyester Resins			
		1.00	Unsaturated polyester (general purpose)	1
		1.00	Unsaturated polyester (general purpose)	2
		1.00	Unsaturated polyester (general purpose)	3
		1.00	Unsaturated polyester (general purpose)	4
		1.00	Unsaturated polyester (general purpose)	5
		1.00	Unsaturated polyester (corrosion-resistant)	6
		1.00	Unsaturated polyester (fire-resistant)	7
	Urea-Formaldehyde Resins			
		1.00	Urea-formaldehyde (molding compound)	1
		1.00	Urea-formaldehyde (syrup)	2
	Vinyl Acetate/Ethylene Copolymer (Latex)			
		1.00	PVAc and vinyl acetate copolymers	4
	Vinyl Chloride/Vinyl Acetate Copolymer			
		1.00	PVC and vinyl chloride copolymers	5
	Vinylidene Chloride/Vinyl Chloride Copolymer			
		1.00	Vinylidene chloride copolymers	1
	Vinylidene Chloride/Ethyl Acrylate/Methyl Methacrylate Termpolymer (Latex)			
		1.00	Vinylidene chloride copolymers	2

Index

There are three indexes for this book. An index of *primary feedstocks and intermediate chemicals* is found on pages 329 through 360. This index cross lists the use of each chemical as a feedstock and as a product in all of the processes contained in the catalog part of the book. The index on pages 360 through 365 does the same for *end products*. The third index which follows locates the more general topics in the text.